Pascal for Electronic Engineers

TUTORIAL GUIDES IN ELECTRONIC ENGINEERING

Series editors
Professor G.G. Bloodworth, *University of York*
Professor A.P. Dorey, *University of Lancaster*
Professor J.K. Fidler, *University of York*

This series is aimed at first- and second-year undergraduate courses. Each text is complete in itself, although linked with others in the series. Where possible, the trend towards a 'systems' approach is acknowledged, but classical fundamental areas of study have not been excluded. Worked examples feature prominently and indicate, where appropriate, a number of approaches to the same problem.

A format providing marginal notes has been adopted to allow the authors to include ideas and material to support the main text. These notes include references to standard mainstream texts and commentary on the applicability of solution methods, aimed particularly at covering points normally found difficult. Graded problems are provided at the end of each chapter, with answers at the end of the book.

1. Transistor Circuit Techniques: discrete and integrated (2nd edition) – G.J. Ritchie
2. Feedback Circuits and Op Amps – D.H. Horrocks
3. Pascal for Electronic Engineers (2nd edition) – J. Attikiouzel
4. Computers and Microprocessors: components and systems (2nd edition) – A.C. Downton
5. Telecommunications Principles (2nd edition) – J.J. O'Reilly
6. Digital Logic Techniques: principles and practice (2nd edition) – T.J. Stonham
7. Transducers and Interfacing: principles and techniques – B.R. Bannister and D.G. Whitehead
8. Signals and Systems: models and behaviour – M.L. Meade and C.R. Dillon
9. Basic Electromagnetism and its Applications – A.J. Compton
10. Electromagnetism for Electronic Engineers – R.G. Carter
11. Power Electronics – D.A. Bradley
12. Semiconductor Devices: how they work – J.J. Sparkes
13. Electronic Components and Technology: engineering applications – S.J. Sangwine
14. Optoelectronics – J. Watson
15. Control Engineering – C.C. Bissell
16. Basic Mathematics for Electronic Engineers: models and applications – J.E. Szymanski
17. Integrated Circuit Design Technology – M.J. Morant

Pascal for Electronic Engineers

Second edition

J. Attikiouzel
Department of Electrical & Electronic Engineering
The University of Western Australia

Chapman and Hall
University and Professional Division

First published in 1984
Reprinted 1985
Second edition 1988

Reprint 1990 published by
Chapman and Hall Ltd
11 New Fetter Lane, London EC4P 4EE

Published in the USA by
Van Nostrand Reinhold
115 Fifth Avenue, New York NY 10003

© 1984, 1988 J. Attikiouzel

Printed in Hong Kong

ISBN 0 412 37590 7
ISSN 0266 2620

This paperback edition is sold subject to the condition that it shall not, by way of trade or otherwise, be lent, resold, hired out, or otherwise circulated without the publisher's prior consent in any form of binding or cover other than that in which it is published and without a similar condition including this condition being imposed on the subsequent purchaser.

All rights reserved. No part of this book may be reprinted or reproduced, or utilized in any form or by any electronic, mechanical or other means, now known or hereafter invented, including photocopying and recording, or in any information storage and retrieval system, without permission in writing from the publisher.

British Library Cataloguing in Publication Data

Attikiouzel, J. (John)
  Pascal for electronic engineers. – 2nd. ed
  1. Computer systems. Programming language
  : Pascal language
  I. Title   II. Series
  005.13′3

  ISBN 0-412-37590-7

Contents

Preface ix
Preface to the Second Edition x

1 Basic Concepts 1
   Algorithm 1
   Programming languages 1
   Software tools 3
   Pascal 4
   Identifiers 5
   Pascal structure 7
   Comments 9
   Examples of bad and good programming 10

2 Scalar Data Type: Constant, Integer, Real. Input-Output 14
   Constant definition 14
   Variable declarations 15
   Integers 16
   Reals 16
   Pascal arithmetic 18
   Arithmetic functions 19
   Input to a program 21
   Output from a program 22
   Formatted output 23
   A step by step development of simple Pascal programs 25
      Case study 1 25
      Case study 2 27

3 Scalar Data Type: Char, Boolean, Enumerated and Subrange. The Array Data Structure 32
   Computer character set 32
   The data type character 33
   Input and output of character variables 34
   Standard function identifiers for character 35
   The data type boolean 37
   Operator hierarchy 38
   Standard functions for boolean 38
   Scalar data type 39
   Enumerated scalar data type 40
   Subrange scalar data type 41
   The array data structure 43

4 Conditional, Repetitive and Goto Statements 47
   Assignment statement 47

	Compound statement	48
	The if statement	48
	The case statement	52
	The while-do statement	54
	The repeat-until statement	58
	The for-statement	61
	The goto statement	67
	Case study 1	69
	Case study 2	70
5	**Functions and Procedures**	**77**
	Why use functions and procedures?	77
	Functions	78
	Local declarations within functions	82
	Scope of identifiers and side effects	86
	Procedures	89
	Procedures with no formal parameters	90
	Procedures with value parameters	90
	Using global variables	93
	Procedures with variable parameters	94
	Procedural and functional parameters	100
	Recursion	102
	Forward directive	103
6	**Structured Data Types: Array, File, Set and Record.**	
	The Pointer Data Type	**107**
	The array structure	107
	Arrays as subprogram parameters	109
	Packed arrays	110
	Strings	111
	The file structure	113
	Standard Pascal procedures for files	115
	Textfiles and standard procedures	116
	The set structure	119
	Set operators	120
	The record structure	123
	Variant record	128
	The pointer data type	131
7	**Case Studies**	**139**
	Network transfer functions	139
	Transfer function analysis program	140
	Active filter synthesis	144
	Active circuit synthesis program	146
	Linear passive circuits	150
	Circuit analysis program	151

Appendix A Syntax diagrams   158

Appendix B Pascal special symbols 162
Standard Pascal identifiers 162
Description of standard functions 162

References 163

Index 165

Preface

In the last few years there has been a tremendous increase in the number of Pascal courses taught at various levels in schools and universities. Also with the advances made in electronics it is possible today for the majority of people to own or have access to a microcomputer which invariably runs BASIC and Pascal. A number of Pascal implementations exist and in the last two years a new Pascal specification has emerged. This specification has now been accepted as the British Standard BS6192 (1982). This standard also forms the technical content of the proposed International Standard ISO7185.

In addition to a separate knowledge of electronic engineering and programming a marriage of engineering and computer science is required. The present method of teaching Pascal in the first year of electronic engineering courses is wasteful. Little, if any, benefit is derived from a course that only teaches Pascal and its use with abstract examples. What is required is continued practice in the use of Pascal to solve meaningful problems in the student's chosen discipline. The purpose of this book is to make the use of standard Pascal (BS6192) as natural a tool in solving engineering problems as possible.

In order to achieve this aim, only problems in or related to electrical and electronic engineering are considered in this book. The many worked examples are of various degrees of difficulty ranging from a simple example to bias a transistor to programs that analyse passive RLC networks or synthesise active circuits. Although it is assumed that the reader has some electronic engineering knowledge, the engineering aspect of each worked example is briefly discussed. No attempt is made to include numerical techniques nor to include examples from all areas of electrical and electronic engineering. What is presented is a number of worked examples to allow most engineering students to become familiar with Pascal in solving their technical problems. This book should serve both the student and the practising engineer as an introduction to standard Pascal, with examples. It should provide also a sound basis for further study of Pascal.

This tutorial guide is primarily intended for first and second year electrical and electronic engineering students. In order to conform with other guides in the series, a number of the more 'advanced' aspects of Pascal have only been mentioned in passing or omitted altogether. A comprehensive list of references to mainstream computer science and electronic engineering books is given to direct the reader in his endeavours for further knowledge.

This tutorial guide was written while I was on sabbatical leave at the University of Essex and I would like to thank my colleagues there who have contributed in many ways to this book. In particular, I am indebted to Gordon J. Ritchie, the author of the first guide in this series, for his continued support and encouragement, and for his painstaking proof reading of numerous manuscripts. I am also grateful to my colleague and office-neighbour at Essex, Peter Jones, who ensured that my Pascal was standard Pascal. Finally, but by no means least, my thanks to Kel Fidler, the consultant editor, for his valuable comments on the manuscript.

Preface to the second edition

In this second edition of the book I have modified all the programs that appeared in the first edition and have added a number of new ones. Chapter 6 has been augmented, and a number of new programs have been added to this chapter. All the programs listed in the book have a typical input-output listing to help the reader.

Since the publication of the first edition I had a number of requests to make the programs available in a machine-readable form. To this end I have all the source programs on a IBM PC 5.25″ diskette which can be obtained from me for a small handling fee.

I would like to thank Dr Greg Crebbin for suggesting some of the new programs and Mr Paul Ostergaard for debugging them.

Basic Concepts 1

☐ To define an algorithm
☐ To compare the different levels of programming languages and introduce software tools.
☐ To introduce the Pascal vocabulary
☐ To write a simple Pascal program.

Objectives

The way to learn programming is to write programs and Pascal is a very good introductory programming language. It is also accepted that problem solving should be learned from the first programming class; but solving programming problems, for their own sake, is usually the last thought that crosses an engineer's mind. Therefore, Pascal in this book is introduced as a tool for solving engineering problems, rather than programming problems. Good programming practice is emphasised throughout this book.

This chapter aims to introduce the basic concepts required to write a Pascal program. It also highlights the need for other software tools necessary to execute and construct a Pascal program.

Algorithm

Whenever computers are involved, problem solving means developing algorithms. Since computers are 'dumb' they require a precise sequence of instructions to perform a given task; such a sequence is called an 'algorithm'. Therefore, an algorithm is a step by step solution to a specified problem.

All algorithms have two things in common, first, the given information to be used (*e.g.* the variables a, b and c of a quadratic expression $ax^2 + bx + c$), and second, a set of permissible operations that can be performed on the information (*e.g.* summation, subtraction, square root and so on).

See Schneider, G.M., Weingart, S.W. and Perlman, D.M. *An Introduction to Programming and Problem Solving with Pascal* (Wiley, 1978. Pages 19–59).

To be an algorithm, a sequence of instructions must possess at least three important properties:
(i) Each instruction must use some of the basic set of operations available in the computer, such as addition, exponentiation etc.
(ii) Produce a result in a specified number of steps and in finite time *i.e.* an algorithm should not loop *forever*.
(iii) Be of finite length.

Programming Languages

Once the steps in solving the problem (*i.e.* the algorithm) have been worked out then a set of instructions or program is required in a language that the computer understands. A digital computer is basically a binary machine, therefore a set of instructions can be written directly in machine language, that is, a sequence of 1s and 0s. While this method of programming may be acceptable for very small programs, it becomes very time-consuming and error-prone as the size and complexity of the program increases.

Program writing is greatly simplified by an 'assembly language' which uses mnemonics such as ADD for addition and SUB for subtraction instead of machine language instructions. Assembly language programs are extensively used with microcomputers; but a programmer requires detailed knowledge of the particular microcomputer being used. In addition, assembly language programs are not portable since each microcomputer has its own assembly language which reflects its own architecture.

> For example an Intel assembly program cannot run on a Motorola microcomputer system.

The solution to many of the difficulties associated with assembly language programs is to use a 'high level' language. High level languages make it possible to program in a more natural language that is closer to the problem than assembly language. Furthermore, structured high level languages such as Pascal impose a programming discipline which makes it easier and faster to write and also to correct and understand programs. Some of the high level languages commonly used by engineers, apart from Pascal, are:

> For further details of assembly language programming, refer to Downton, A. C. *Computers and Microprocessors: components and systems* (Van Nostrand Reinhold (UK), 1984), Chpater 7.

FORTRAN (**FOR**mula **TRAN**slator) is the first (1954) widely used high level language and is mainly used for scientific applications.

ALGOL 60 (**ALGO**rithmic Language) was developed in 1958 and it is a block structured algebraic language extensively used in Europe for numerically oriented problems. A number of extended versions are presently being used.

APL (**A P**rogramming **L**anguage) was developed (1967) primarily for efficient handling of mathematical functions. It is widely used in engineering and mathematics.

> For those of you that know BASIC see: Brown, P.J. *Pascal from BASIC* (Addison Wesley, 1982).

BASIC (**B**eginner's **A**ll-purpose **S**ymbolic **I**nstruction **C**ode) is not a sophisticated language and is easy to learn. Because it is usually interpreted at run time instead of being compiled it is very interactive with the programmer. BASIC was developed in 1965.

COBOL (**CO**mmon **B**usiness **O**riented **L**anguage) was developed (1960) mainly for accounting and business applications.

C is a structured high level language designed in 1978 for efficient control of machine resources while making machine independent programming possible. It is widely accepted as a 'portable assembler'.

Worked Example 1.1

Write the necessary instructions in machine language, assembly language and a high level language to add 5 to a variable.

> You are not expected to be able to write in all three languages given here, but only to appreciate the difference in level.

Solution:

Machine language	Assembly	High Level Language
0111 1000	MOV A,B	B: = B + 5;
1100 0110	ADI 05	
0000 0101		
0010 0111	MOV B,A	

It can be seen from the above solution that writing in a high level language is more natural and that it requires the minimum number of instructions.

Software Tools

It has been mentioned that the computer is a binary machine, that is, the information the hardware requires at the processing stage must be in binary form. A Pascal source program or, for that matter, any other high level language program, cannot be directly executed by the computer. The task of translating Pascal programs and generating binary form instructions is performed by a program called a 'compiler'. It gets its name because its function is to compile a list of machine instructions; it could also be called a 'translator'. The input to a compiler is the whole source program and the output is an object program and a 'listing' program. The 'listing' is a listing of the source program with line numbers, together with any error messages.

In practice, many widely different compilers are available for the same high level language (called 'implementations'); for example, I am aware of over 30 different implementations of FORTRAN. The difference forced on various compilers is mainly due to computer architecture. Therefore some aspects of a language may be implementation dependent, which either enhance or diminish the capabilities of a particular compiler. Therefore, before you start coding a program you must ensure that you are aware of the implementation that is available for your use.

In translating the source program the compiler checks for any errors which the programmer may have made in writing his source program. These errors are referred to as 'compile-time' errors and usually a good compiler indicates their nature and approximate position. These errors are also known as 'syntax' errors. If the program is free of syntax errors then the object code is executed or 'run'. During this step the computer sequentially executes the program and if any further errors are detected, then execution of the program stops and the error messages are printed. Such errors are, for example, the calculation of the logarithm of -1 or a division by zero. These errors are known as 'run-time' errors and are violations of the Pascal internal design rules.

> Error messages are implementation defined. See your user manual for more details.

> Run time errors, also called semantic errors, can also be caused by the limited word manipulation capability of your computer.

Another useful program that is needed in the creation and modification of the source program is a text 'editor'. Editor commands are numerous and depend on the implementation. Among the most useful commands are: insert, modify or reposition characters and add or delete characters or lines from the source program. When the source program has been inserted into the computer, using a keyboard, it is usually saved as a file on the computer's auxiliary memory such as a floppy disk or magnetic tape. It can be retrieved from there for further processing whenever it is required.

Fig. 1.1 shows a flow diagram of the process of writing, compiling and running a program. It can be seen from this diagram that the editor is a major tool for correcting or 'debugging' a source program; therefore familiarity with a particular editor is of utmost importance. Before attempting to write any program in Pascal or in any other language make sure that:

(i) You are familiar with the editor that is available for your use.
(ii) You know the necessary operating system instructions to compile and execute your program.
(iii) You are aware of which language implementation your establishment supports.

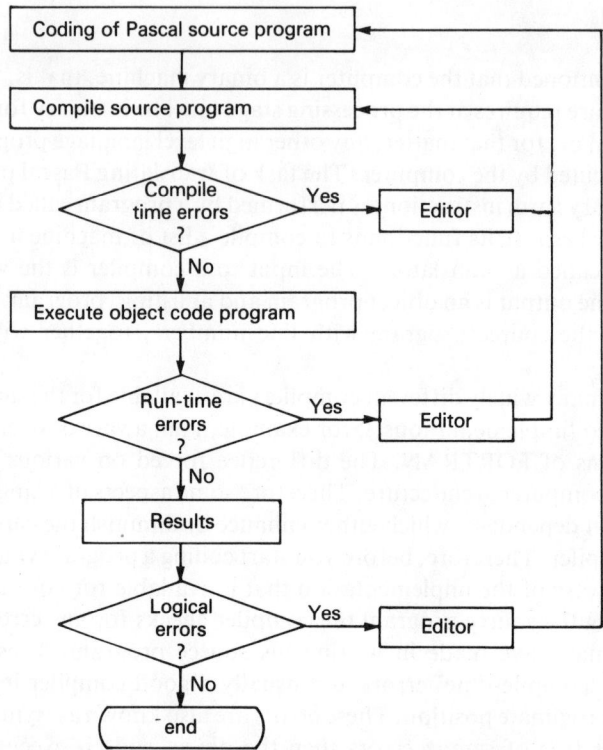

Logical errors are flaws in the algorithm or its expression as a program.

Fig. 1.1   Flow diagram of program construction.

Pascal

Pascal was named after the mathematician Blaise Pascal and is a descendent of ALGOL 60. The language was developed in Switzerland by Professor Niklaus Wirth to allow the development of well-structured and well-organised programs. Pascal is a general purpose language that is applicable to numerical and non numerical type problems.

See Jensen, K and Wirth, N. *Pascal: User Manual and Report* (Springer-Verlag, 1975).

Pascal programs can process many different kinds of data using a number of arithmetic and logical operations. Like many other high level languages Pascal uses the 'simple' (or 'primitive' or 'unstructured') data types: real, integer, boolean and character but in addition users are able to create their own data type. The most important contribution of the Pascal language is the concept of 'data type'. A data type is described by a set of rules that specify the format and operations for the elements of that type. The basic subdivision of data types as shown in Fig. 1.2 is

Here an overview is given. The details of the various data types are discussed in the following chapters.

(i) 'scalar data type' which are either Pascal-defined or user-defined. These are: *real, integer, boolean* and *char* data types.
(ii) 'structured data type' which specify the way in which simple data types are combined and their interrelationship. These are of type: **array, file, record** and **set**.
(iii) 'pointers' which define dynamic variables.

More details about the various data types are given as they are encountered in the text.

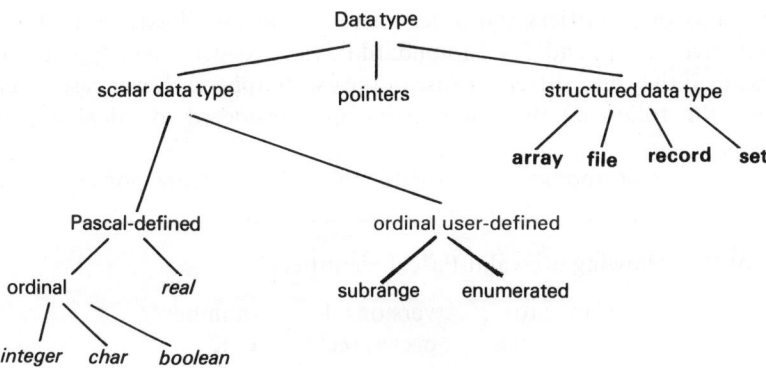

Fig. 1.2 Pascal data types.

Ordinal types are *char*, *integer* and *boolean*. More details in Chapters 2 and 3.

All languages, whether human or computer programming, make use of a vocabulary. The Pascal vocabulary consists of letters, digits and special symbols. Instructions are constructed out of this vocabulary according to predefined Pascal syntax rules.

A letter may be any one of the 26 letters of the English alphabet in either upper-case or lower-case. A digit is any one of the ten Arabic numerals 0–9. These digits can form numbers which can be represented as integer numbers (such as 1, 567, 10987) or as real numbers in fixed point format (1000.0, 0.95, −87.29) or in scientific notation (1E3, 9.5E−1, −8.729E+1).

The recently accepted standard Pascal supports only upper case.

E is read as *times ten to the power of*.

The number of special symbols used in Pascal is quite high (60) and can be divided into two categories: (i) punctuation and algebraic symbols and (ii) word symbols or reserved words. These are listed in Table 1.1. Throughout this book reserved words are printed in **bold** face for ease of learning. Your computer terminal may have only one type-face.

Table 1.1 Pascal Special Symbols

Punctuation and Algebraic symbols				Reserved word symbols				
+	−	*	/	and	downto	if	or	then
>	<	[	]	array	else	in	packed	to
;	,	:	.	begin	end	label	procedure	type
^	(	)	..	case	file	mod	program	until
<>	<=	>=	:=	const	for	nil	record	var
=	{	}		div	function	not	repeat	while
				do	goto	of	set	with

There are two symbols in the table that you do not see: space and carriage return.

Identifiers

The various objects used in a Pascal program are given *mnemonic* or *symbolic* names by which they can be identified. Good programmers choose identifier names that are descriptive of the identifier's meaning or function. For example there is no point in using TIME as an identifier for voltage. Pascal identifiers are composed of letters and digits and can be of any length with the overriding restriction that the first character must be a letter. The special symbols given in Table 1.1 cannot be used as identifiers.

Identifiers may be of any length but some Pascal implementations limit the recognizable identifier length to a total of eight letters and digits.

> See next section for more details about syntax diagrams.

The syntax of identifiers and other Pascal syntax are illustrated by the *syntax diagrams* given in Appendix A. In standard Pascal two identifiers are considered to be the same if they only differ because of the use of upper or lower case letters. For example, the following three identifiers are considered identical in standard Pascal:

      StepResponse   STEPRESPONSE   stepresponse

Worked Example 1.2 Which of the following are valid Pascal identifiers?

Capacitor	version1.1	-number
volts	cycles/sec	U.K
$100	*sec	Feb83
2ndvalue	msec	14Feb84

Try this, before checking the Solution at the bottom of the page.

Every implementation of Pascal has a number of predefined identifiers. These are called *standard identifiers* and are used to define the various data types and also to facilitate the calculation of a number of operations such as trigonometric and arithmetic operations. The identifiers included in every Pascal implementation are shown in Table 1.2. Your particular implementation may have additional ones. In this book standard identifiers only will be used in writing the various programs. If a Pascal program is written with only standard identifiers then the program is portable, *i.e.* it can run in another establishment that supports standard Pascal. Standard identifiers, in contrast with reserved words, may be redefined by the programmer but this practice is definitely not recommended. Standard Pascal identifiers are printed in *italic* for ease of identification throughout this book.

Table 1.2 Standard Pascal Identifiers

> Ignore the headings; these are explained later.

Files	Constants	Types	Procedures		Functions		
input	*false*	*boolean*	*get*	*readln*	*abs*	*odd*	*arctan*
output	*true*	*integer*	*new*	*reset*	*chr*	*ord*	*succ*
	maxint	*real*	*pack*	*rewrite*	*cos*	*pred*	*trunc*
		char	*page*	*unpack*	*eof*	*sin*	*eoln*
		text	*put*	*write*	*exp*	*sqr*	*round*
			read	*writeln*	*ln*	*sqrt*	

Reserved words and identifiers should not be confused. Reserved words have a fixed meaning and can not be redefined in a program; they can be considered as the basic tools necessary to write a Pascal program. Standard identifiers help the programmer with various operations such as arithmetic and logical functions or with file handling. Standard identifiers can be redefined, but if this is done a valuable aid is lost. It is possible to write Pascal programs without standard identifiers but it is impossible to do so without reserved words.

Identifiers introduced by the programmer are called 'user-identifiers' or simply identifiers. These are mnemonic names used to represent constants, various data

Solution to Worked Example 1.2: Only Capacitor, volts, msec and Feb83 are valid Pascal identifiers.

types, variables, procedures and functions. More details about these are given in the following chapters.

Which of the following are valid Pascal identifiers? **Worked Example 1.3**

Temperature	Table1.2	(black)
t	50 hz	gain
true	A + B	ω
root	phase-shift	inductor
squareroot	next char	e(3)
log	array	omega

Solution: The identifiers in the left hand column and gain, inductor and omega in the right hand column are correct.
N.B. Redeclaring true, although correct, should not be done since it is a pre-defined standard identifier.

Pascal Structure

As with any natural language, every programming language has a set of rules associated with it. This usually strict set of rules describes how a valid program may be constructed in the language. Pascal is a language specifically designed to facilitate structured programming. Three key elements contribute to making Pascal a structured language: (i) declarations, (ii) structured data and control, and (iii) subprograms *i.e.* procedures and functions.

A Pascal program consists of a program HEADING and a BLOCK followed by a period (full-stop) as shown in Fig. 1.3. Quantities within rectangular boxes require further definition and quantities within round ended boxes are actual characters which must appear within the program text and require no further definition.

See also Appendix A.

Fig. 1.3   Syntax diagram of Pascal programs.

The language semantics can also be expressed using Extended Backus Naur Form (EBNF) as:
  program = heading block "."
Symbols in double quotes are equivalent to symbols in round ended boxes. The BS6192 Pascal standard is defined using EBNF form.

The overall Pascal structure can be subdivided further as shown in Figs. 1.4 and 1.5. The program HEADING simply gives the program a name and specifies its operating environment (input and/or output). For example, in a batch system input might be a card reader and output a line printer; in an interactive system input might be the keyboard and output the display screen. A typical program HEADING would be

  program Test (*input,output*);

The reserved word **program** indicates the beginning of the program. The identifier Test is the chosen name for the program. By specifying both *input* and *output* it

program is a reserved word and is unique in the program. *input, output* are standard identifiers and are also unique in the

7

program. Notice the use of **bold** and *italic* for reserve words and standard identifier. You do not need these type faces to write your program. They are only used here to help with your learning.

is indicated that the program will have both input and output data from and to the standard input and output devices which are attached to the system being used *i.e.* a terminal and/or printer.

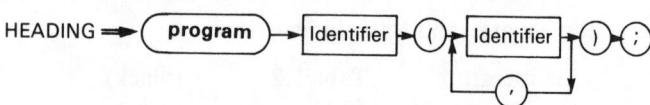

Fig. 1.4   HEADING syntax diagram.

The BLOCK is divided up into two parts as shown in Fig. 1.5: (i) the definition and declaration part and (ii) the main body of the program which consists of a number of statements. All parts of the BLOCK are optional except the statement part. It is important to remember that the precise role of the various identifiers in the program must be defined and declared at the beginning of the program. In a Pascal program a variable is created by a declaration. To be able to declare or define a variable it is necessary to specify both its type and its name; this is shown in Chapters 2 and 3. Furthermore this definition and declaration must be done in the prescribed order shown in Fig. 1.5. The meaning of the definitions (**L**abel, **C**onstant, **T**ype) and declarations (**V**ariable, **S**ubprograms) is described fully in the following Chapters.

In Pascal there are two types of subprograms : **procedure** and **function**.

The statement part of a program always starts with **begin** and terminates with an **end**. The statements in the program manipulate the declared data types. Successive statements are separated by a semicolon. Please remember that the semicolon is a separator and not a terminator.

Note the := means becomes equal to.

A simple statement performs a single action. For example A: = B + 5; is a statement that adds 5 to a variable called B and assigns the result to a variable called A. An instruction which alters the value of a variable is called an 'assignment statement'. The assignment statement is the most frequently used statement in Pascal. Note that when a value is assigned to a variable, its previous value is lost.

Definitions and declarations convey information to the compiler about the type of data and about data values; whereas statements specify actions to be taken by the computer when the program is being executed. 'Statements' in this context can be considered as synonymous to 'commands' or 'instructions'.

A group of statements grouped together by **begin** and **end** is called a 'compound statement'. The compound statement specifies that its component statements be executed in the same order as they are written. An example for a compound statement is:

begin
   a: = a + 5;
   c: = c * c + a
end;

begin and **end** are best regarded as flags to the compiler, telling it where blocks of code start and where they terminate. They can also be considered as a left parenthesis (**begin**) and a right parenthesis (**end**) surrounding a set of statements. This is one of the features that makes Pascal so useful in writing well-structured programs.

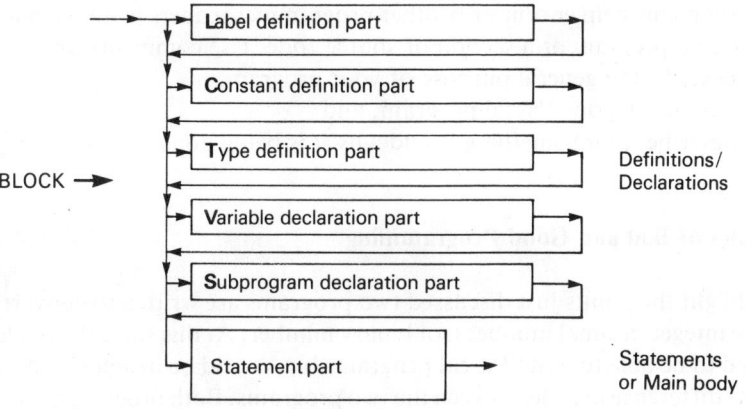

Fig. 1.5 BLOCK syntax diagram.

To help remember the order **LCTVS** : **L**ight **C**omedy **T** – **V** **S**how.

A typical Pascal program layout looks like this:

```
program Test (input,output);              HEADING
label     ....
const     ....                            DEFINITIONS
type      ....
var       ....                            DECLARATIONS
procedure
function

begin
    statement
    statement                             STATEMENTS
    begin
        statements
    end
end.
```

Note that the statement part of a Pascal program has the form of a compound statement.

The last . in a program is unique.

The layout of the program text is of no concern to the compiler *i.e.* Pascal statements are completely free-format and may appear anywhere on a line. But consecutive words and numbers must be separated *i.e.* the statement

is different from   begin ing: = th end;
                      begining: = thend;

Any number of spaces and/or newlines may be used to separate identifiers and special symbols used in the program. More than one statement may be placed on a single line as long as they are separated by a semicolon.

Layout of the source program should be designed for ease of reading and understanding. Good habits of indentation should be adopted to enhance the readability of your program and to highlight its logical structure.

Have a look at the book by Ledgard, H.F., Hueras, J.F. and Nagin, P.A. *Pascal with style* (Hayden, 1979) regarding program style. This book also includes a useful program *prettyprinter* which automatically indents and *pretties* your programs.

I prefer (∗ and ∗) instead { and } because it stands out better. If both symbols are available then it is a matter of personal choice.

Comments

Explanations about the operation of a program can be given as 'comments'. Comments are placed between curly brackets {.........} or between (∗.........∗) that is to say that { is equivalent to (∗ and } is equivalent to ∗). Comments may be placed anywhere that a space or end of line may appear. There is no need for a semicolon at the end of a comment since the compiler ignores all comments. A comment is not an executable statement but is there for the benefit of the human reader.

This might include you, six months after you have written the program.

Good comments help enormously other programmer readers in understanding the purpose of a program or a section of source code. Use comments to:
  (i) describe the general purpose of your program,
  (ii) paragraph your Pascal program, and
  (iii) describe what something is and why it is being done.

Examples of Bad and Good Programming

To highlight the points just discussed two programs are written to convert a given positive integer decimal number to a binary number. At this stage the reader is not expected to be able to write Pascal programs but should be in a position to appreciate the difference in style between the two programs. Both programs, if executed, give the correct results. The algorithm used in both programs is:

Before you code any program make sure that you have first developed a correct and efficient algorithm.

Step 1: Divide the decimal number by 2. Save the remainder *i.e.* the modulo.
Step 2: If the quotient is zero, then proceed to step 3. If the quotient is not zero, replace the number with the quotient and repeat step 1.
Step 3: The binary representation of the decimal number is the remainder, starting with the first remainder in the least significant position.

Program 1

```
program    decimaltobinary (input,output);          const    base=2;var
decnumber,quotient, remainder:integer; begin
write('Enter   positive decimal integer = ');readln(decnumber);
write('The binary equivalent in reverse order   is  = ');
if     decnumber=0   then write('0')    else    begin    repeat begin
quotient:=decnumber       div      base; remainder:=decnumber       mod
base;write(remainder);    decnumber:=quotient; end             until
quotient=0;end;end.

Execution begins...

Enter   positive decimal integer = 56
The binary equivalent in reverse order   is   =    0 0 0 1 1 1

Execution terminated.
```

Since a Pascal program can be written in free format the above program is correct, but its logical structure is not obvious nor is the program clearly understood. By using indentation and adding comments the program can be rewritten as:

Program 2

All programs written in this book can be made acceptable to other Pascal implementations which either accept lower case or upper case only, since no two identifiers are distinguished in programs by a change to upper or lower case.

    program DecimaltoBinary(*input,output*);
    (* This program converts a given decimal number to
       binary number. N.B. The binary digits are printed
       in reverse order.   *)
    const base = 2;
    var   decnumber,quotient,remainder : *integer*;
    (* Statement part starts from here *)
    begin
      write ('Enter positive decimal integer = ');*readln* (decnumber);

```
              write ('The binary equivalent in reverse order is = ');
              if decnumber = 0 then write ('0')
                            else
                                begin
                                  repeat
                                    begin
                                      quotient: = decnumber div base;
                                      remainder: = decnumber mod base;
                                      write (remainder);
                                      decnumber: = quotient
                                    end
                                  until quotient = 0
                                end
end.
```

The algorithm used in this example is not the best which can be used for converting a decimal number to binary but it serves the purpose of highlighting the correspondence between an algorithm and a program. A better algorithm is used for arithmetic conversions in Chapter 4 (Example 4.4).

From the above listing we can identify the first line of the program as being the program-heading. This contains the name of the program: DecimaltoBinary, and it has also specified that both input and output are used. The program heading is followed by three lines of comment describing the program. Only one constant is defined together with three variable identifiers in the definition/declaration part of the program. The main body of the program follows the next comment and starts with **begin** and terminates with the last **end**. The whole program is then terminated with a period.

The above program is photographically reproduced in Fig. 1.6 as it was printed on a dot-matrix printer. Your program will be in this form *i.e.* it does not contain **bold** and *italic* print faces.

```
program DecimaltoBinary(input,output);
(* This program converts a given decimal number to
   binary number. N.B. The binary digits are printed
   in reverse order.                                    *)

const base = 2;
var    decnumber,quotient,remainder : integer;
(* Statement part starts from here*)
begin
   write ('enter positive decimal integer = ');readln(decnumber);
   write('The binary equivalent in reverse order is = ');
   if decnumber = 0 then write ('0')
                else
                    repeat
                       quotient:=decnumber div base;
                       remainder:=decnumber mod base;
                       write(remainder);
                       decnumber:=quotient
                    until quotient = 0;
   writeln
end.

Execution begins...

enter positive decimal integer = 56
The binary equivalent in reverse order is =   0 0 0 1 1 1

Execution terminated.
```

Fig. 1.6   Program as listed on a printer.

I suggest at this point that you have a glance at the programs given in Chapter 7. It might help to get a better picture of a Pascal program.

Summary

In this chapter it has been seen that it is simpler to write programs in high level language rather than in assembly language or machine code. It was pointed out that a knowledge of an editor program is necessary for the purpose of inserting and subsequently modifying programs in a computer. The compiler with the aid of error messages indicates the syntax errors. Run time errors occur when the program is being executed and are caused by misuse of the Pascal language. All these errors can be corrected with the use of the editor program. It has also been pointed out that an awareness of which Pascal implementation an establishment supports is useful since this has an effect on program writing.

The most important part of programming is the design and development of correct and efficient algorithms. An algorithm is a step-by-step process for solving a problem and must possess the properties discussed in this chapter.

Pascal, like all other computer languages, relies on a vocabulary and certain rules which must be obeyed for the program to work. Pascal is a well structured language and one of its most important advantages over other high level languages is the concept of data type.

Pascal programs are divided into three main parts
(i) The program heading, which names the program and specifies its input and output environment.
(ii) The Block, which consists of the definitions and declaration part, where the type of all constants and variables are defined and declared, and the statement part which contains the instructions of the program.
(iii) A period which indicates the end of the program.

All variables used in a Pascal program must be declared in the specified order **(L,C,T,V,S)**. Pascal programs should be written with comments and proper indentation to help the program reader with the understanding of the program.

Problems

1.1 Develop an algorithm to find the roots of quadratic equations of the form $ax^2 + bx + c = 0$.

1.2 Define: (i) Semantic errors, (ii) Syntax errors and (iii) Logical errors.

1.3 (i) Specify the key elements that constitute a Pascal program.
    (ii) Is the definition and declaration order important? State this order.

1.4 Classify each of the following:

begin	1E1	trunc
character	atan	1.3656
2487986	maxint	succ
end	13,000	boolean
data	real	repeat
+	input	not

1.5 Describe the basic subdivision of data types.

1.6 Which of the following are valid Pascal identifiers:

| program1 | program | condition | constant |
| 3rdvalue | step-resp | OpAmp | exit |

99	199	Unix1.1	ReSiStOr
div3	div 3	*a	(* Comment *)
phase shift	integer	exp	reals
tan	TAN	atan	sine
log	cos	SUCC	testrun

1.7 Compare (i) a compiler (ii) an interpreter and (iii) an assembler.

1.8 List the most important commands of the editor that is available for your use. (*Hint* refer to your manual).

1.9 Why are comments important? Can you think of a reason why there are two separate symbols for comments?

1.10 Develop a better algorithm than the one given in the text to convert a decimal integer number to a binary number.

2 Scalar Data Type: Constant, Integer, Real.
Input – Output

Objectives
- ☐ To introduce the concepts of constants and variables within a Pascal program.
- ☐ To define the data types *real* and *integer* and specify their limitations.
- ☐ To introduce procedures for input and output.
- ☐ To find the output format limitations of a Pascal compiler.
- ☐ To write a number of simple Pascal programs.

Variables need not be declared in BASIC or FORTRAN.

All identifiers used in a Pascal program must be defined or declared at the beginning of the BLOCK. The order of definition/declaration (**L,C,T,V,S**) is very important; syntax diagrams for these are given in Appendix A. All identifiers used in a program must be declared. The **const** definition is used for parameters that do not change their value throughout the execution of the program. The **var** declaration can be any one of the data type: *real, integer, boolean* or *char* (short for character). In addition to these four simple types of variables, Pascal allows more complex variable types known as 'structured' data types. These include **array, record, set** and **file** and are described in Chapter 6.

In this chapter the **const** declaration and two types of **var** declarations, *real* and *integer*, are introduced together with simple input and output procedures. Data type *char* and *boolean* are discussed in Chapter 3.

Constant Definition

In any problem solving routine there are basically two kinds of data that are used: constants and variables. The constant's value remains unchanged throughout the program, for example, the value of pi. On the other hand variables have different values assigned to them at different stages during the execution of the program.

In Pascal the constant definition part starts with the reserved word **const** followed by each constant's identifier, an equal sign and finally its value *i.e.* a **const** definition looks like an equation. A semicolon separates successive definitions. For example:

Note an equal sign (=) is used for constant definitions.

  const pi = 3.1415926536;   (∗ Real type ∗)
      InputResistance = 2.2E6;
      RAM = 512;   hr = 12;   (∗ Integer type ∗)
      Fred = 'F';   asterisk = ' ∗ '; (∗ Character type ∗)
      fail = *false*;   (∗ Boolean type ∗)
      note = 'Transistor is saturated'; (∗ String type ∗)

The type of the constant is recognised by the compiler as follows:

(i) Real constants must have digits on either side of the decimal point or an E. E is pronounced '10 to the power of'.

(ii) Integer constants must not have fractional or exponent parts. A decimal point is not allowed.

(iii) Character constants must be a single character enclosed in quotes.

(iv) Boolean constants must be either set to *true* or *false*.

true and *false* are standard Pascal identifiers.

(v) String constants must contain two or more characters enclosed in quotes.

No attempt should be made to change the value of a constant within the program otherwise an error is flagged.

Variable Declarations

In Pascal, the variables are named by identifiers and their type is declared in the **var** declaration section of the program. Declaration of all variables begins with the reserved word **var** followed by a list of identifiers together with their type specification. Type specification aids the compiler in making error checks, such as type mismatch. The list of identifiers and their type are separated by a colon. The syntax diagrams for the different **var** declarations are given in Appendix A. For example:

The order within a **var** declaration is not important.

```
var i,j,k    : integer;   (* i,j and k are declared as integers *)
    freq,dB  : real;      (* freq and dB are declared as reals *)
    initial  : char;      (* initial is declared as a character *)
    test     : boolean;   (* test is declared as boolean *)
```

Several identifiers may be declared in the same statement, separated by a comma, if and only if they have the same type specification. This is shown in the example above in the declaration of *real* and *integer*.

Make sure that you notice the position of the colon (:) relative to the identifier and its type and that it is a colon and not a 'becomes equal' sign (:=) nor an equal sign (=).

Watch this trap.

Which of the following are valid Pascal declarations?

Worked Example 2.1

const twopi = 6.28;	**const** twopi = 6.28;	**const** twopi = 6.28;
var cap,IND,Res:*real*;	**var** cap :*real*;	**var** correct :*boolean*;
number :*integer*;	IND,Res :*real*;	cap :*real*;
correct : *boolean*;	number:*integer*;	number : *integer*;
	correct : *boolean*;	IND : *real*;
		Res : *real*;

Solution: All are valid. Within a **var** declaration the order is not important. Identifiers of the same type can be declared as a group or individually. In fact all three sets of declarations are the same.

Worked Example 2.2 Correct the following **var** declaration statements.

$\quad$ $a.$**var** trnstr,beta : *real*; $\quad$ $b.$**var** filter,response = *real*;
$\quad\quad\quad$ i;j $\quad\quad\quad$: *integer*; $\quad\quad\quad$ timestep : *integer*;
$\quad\quad\quad$ time $\quad\quad$:= *real*; $\quad\quad\quad$ response : *integer*;

Solution:
$a.$**var** trnstr,beta : *real*; $\quad$ $b.$**var** filter,response : *real*;
$\quad\quad$ i, j $\quad\quad\quad$: *integer*; $\quad\quad\quad$ timestep : *integer*;
$\quad\quad$ time $\quad\quad$: *real*; $\quad\quad\quad$ response:*integer*; is not possible
$\quad\quad\quad\quad\quad\quad\quad\quad\quad\quad\quad\quad\quad\quad\quad\quad\quad\quad\quad$ since response has already
$\quad\quad\quad\quad\quad\quad\quad\quad\quad\quad\quad\quad\quad\quad\quad\quad\quad\quad\quad$ been declared as *real*.

Integers

> Word length is the maximum number of binary digits that a computer can operate on simultaneously.

The *integer* type in Pascal permits the representation of integers (whole numbers) in the computer. Theoretically there could be integer values from $-\infty$ through 0 to $+\infty$, but computers have finite word lengths usually of 8, 16, 32 or 64 bits. The largest integer number which a computer can represent is $(2^{n-1} - 1)$, where n is the word length. For example, in microcomputers with 16 bit word length the maximum integer value that can be represented is 32767. In Pascal an integer is represented by a string of digits with an optional sign inserted to the left of the most significant digit.

In order to avoid having to know this limit, Pascal has a predefined constant identifier *maxint* (see Appendix B), whose value is the largest integer number that can be held in the computer. The range of integers that can be handled is therefore implementation dependent and can be represented as shown in Fig. 2.1.

$\quad -(maxint+1) \quad \ldots\ldots\ldots \quad -2\ -1\ 0\ +1\ +2 \quad \ldots\ldots\ldots \quad +maxint$

Fig. 2.1 Integer number representation.

If the permissible range is exceeded during the execution of a program then an overflow error occurs and the program stops executing.

Reals

> There is generally only one digit before the decimal point in floating point notation.

Real variables are represented in a computer in a different form than integers to allow a greater range of values. There are two ways of writing real values in a Pascal program: (i) fixed point decimal notation, and (ii) scientific or floating point notation. An optional sign can be inserted to the left of the first digit in either notation. The following three requirements must be remembered when assigning numeric values to real variables or when representing real data:

  (i) The magnitude of a real variable must begin with a digit.
  (ii) If a decimal point is used in the scientific notation then the number must include a digit on both sides of the decimal point.
  (iii) In scientific notation the exponent must be an integer number.

Examples of valid and invalid real are:

Decimal notation	Scientific notation	
2.7894	0E0	—
+2.784592	−3E7	are valid
−328.12	3E−3	real
0.5	3.0E+21	
5.8	−6.0E−8	—
−.5	3.E−6	—
3.	−2.7E3.0	are invalid
.002	1,000E10	reals
5.	.1E3	—

Floating point constants always represent real numbers even if the decimal point is missing.

Correct the following real and integer data:

Worked Example 2.3

real	integer
1.	3.0
−.2	1,000
−5.E−6	1E1
1,000.00	1.00
−5.5E+6.0	0.1

Solution:

1.0	3
−0.2	1000
−5.0E−6	10
1000.00	1
−5.5E6	not possible

As in the case of integers, owing to finite word length, reals have a restricted range of values. With real numbers two limits exist: one is in the representation of large numbers and another in the representation of very small numbers. Assume that real numbers can be represented as shown in Fig. 2.2.

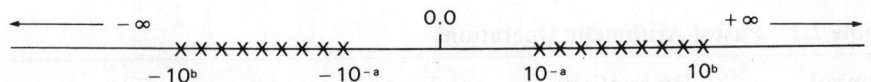

Fig. 2.2 Real number representation.

The cross hatched areas represent the regions where real numbers can be represented in a computer. In the other areas the numbers are either too small or too large. The following limits apply to three widely used computers:

KA 10	$a = 31$	$b = 38$
VAX 11/750	$a = 38$	$b = 38$
CDC 6000	$a = 294$	$b = 322$

A program is given in Chapter 4 to find the range of *real* in your system.

17

Representation of real numbers in the computer is not exact and a very important point to consider with reals is precision *i.e.* the maximum number of significant digits a real number can have in the computer. This limit varies from computer to computer and can be as small as 6 or as many as 18 decimal digits. Limited precision forces numbers with too many significant digits to be rounded off or, even worse, to be truncated. For example assume that the number 0.77777777 . . . is to be represented in a computer with only 6 decimal digit accuracy, then by rounding to ten decimal places the error is − 0.0000002222, and by truncation it is 0.0000007777. Unfortunately these errors accumulate and can give erroneous results, especially if numerical methods are used to solve scientific problems. The programmer should be aware of these computational errors.

Numerical methods exist to minimize these problems. As a rule avoid subtraction of almost equal numbers and avoid testing real values for equality.

> Atkinson, L.V. and Harley, P.J. *An Introduction to Numerical Methods with Pascal* (Addison-Wesley, 1983) is a very good book for those intending to use numerical methods.
>
> If you must use real values and x, y are declared as *real* then use abs(x − y) < = epsilon, where epsilon is a very small number *e.g.* 1E − 9. If you do not know the magnitudes of x and y use abs(x − y) < = x * epsilon.

Pascal Arithmetic

There are a number of arithmetic operations that can be applied to *real* and *integer* numbers. These are summarised in Table 2.1. Of utmost importance is the type of result that is obtained from each arithmetic operation.

Pascal permits the construction of arithmetic expressions of arbitrary complexity and length. As with other high level languages multiplications and divisions have precedence over addition and subtraction.

Pascal uses the reserved words **div** and **mod** to perform divisions with only integer numbers. **div** retains the integer part of the quotient and discards any fractional part. **mod** retains the remainder after an integer division. For example:

> **div** is division by truncation.

$$7 \; \textbf{div} \; 3 = 2$$
$$7 \; \textbf{mod} \; 3 = 1$$
$$-7 \; \textbf{div} \; 3 = -2$$

div and **mod** can not be used with real values, if such an attempt is made the compiler flags an error. An error also is flagged if an attempt is made to compute the **mod** with a negative or zero divisor.

> For some Pascal implementations i **mod** j with j zero or negative does not cause an error but gives erroneous results.

If an arithmetic expression contains both integer and real variables then the type of result for this expression is real. A slash / can be used to divide two integer numbers but the result is real.

Table 2.1 Pascal Arithmetic Operations

Symbol	Description	Type of operand	Type of result
+	Addition	real	real
		integer	integer
−	Subtraction	real	real
		integer	integer
*	Multiplication	real	real
		integer	integer
/	Division	real or integer	real
div	Integer division	integer	integer
mod	Modulo	integer	integer

If an expression contains more than one operator of the same order of precedence, then evaluation proceeds from left to right. Subexpressions in parentheses are evaluated first. Although the hierarchy is well defined parentheses should always be used whenever the order of execution is not obvious. The hierarchy is:

(i) Parenthesised expressions.
(ii) * / **div** **mod**
(iii) + −

Parentheses improve readability of your program, use them whenever possible.

Points to note:
(i) Make sure you distinguish between **div** and /
(ii) If an expression contains both real and integer numbers the result is always real.
(iii) You cannot use real data types with **div** and **mod**.

Evaluate the following expressions and specify the type of result:

Worked Example 2.5

a. 9 / 3
b. 9 **div** 3
c. 5 **div** 1.0
d. 5 + 12/12
e. 12.5 − 12.5/2.0
f. 3 * (5.5 − 4.5) **div** 2
g. 3 + 2 * 4 − 33 **div** 4 − 3
h. 4.7/2 + 15 * (5.1 + 6.2)
i. 25.5 **mod** 25.5
j. 9 **mod** 3
k. 9 **mod** −3
l. (7 **mod** 2) **div** 1

Solution:
a. 3.0 real
b. 3 integer
c. type mismatch
d. 6.0 real
e. 12.5 − 6.25 = 6.25 real
f. type mismatch
g. 3 + 8 − 8 − 3 = 0 integer
h. 171.85 real
i. not valid
j. 0 integer
k. not valid
l. 1 integer

Care should be taken when performing arithmetic operations not to produce results outside the permissible range for integers and reals *i.e.* try to avoid overflows and underflows. Another very common mistake when dealing with arithmetic operations is to try to divide by zero.

Arithmetic Functions

Arithmetic using a Pascal program is further enhanced with the use of a number of predefined standard identifier functions (see Appendix B). The functions can be used by making a 'function call' which consists of naming the required function followed by the function's argument in parentheses. The argument could be a number, a variable or even an expression *e.g. ln*(0.3536), *sin*(x), *sqrt*(b * b − 4 * a * c). Special attention must be paid to the type of argument that can be used and, most important, to the type of result that is obtained. The standard Pascal arithmetic functions for reals and integers are shown in Table 2.2.

Table 2.2 Arithmetic Standard Identifier Functions

Function name	Description	Type of argument	Type of result
abs	absolute value	real	real
		integer	integer
arctan	arc tangent	real or integer	real
cos	cosine	real or integer	real
exp	exponential	real or integer	real
ln	natural log	real or integer	real
pred	predecessor	integer	integer
round	rounding	real	integer
sin	sine	real or integer	real
sqr	square	integer	integer
		real	real
sqrt	square root	real or integer	real
succ	successor	integer	integer
trunc	truncation	real	integer

Note only a t difference between *sqr* and *sqrt*.

Consider a number of these functions:

sin(a),*cos*(a), *arctan*(a)	The argument, *a*, may be real or integer and it must be given in radians.
ln(a)	If the argument is non-positive a run-time error is flagged.
sqrt(a)	If the argument is negative a run-time error is flagged.
trunc(r)	This function converts a real number to an integer number by truncating the decimal part.
round(r)	This function also converts a real number to an integer by rounding to the nearest integer.
succ(i)	This function produces the successor of the integer number i *i.e.* if i = 5, *succ*(i) returns 6.
pred(i)	This function produces the predecessor of the integer number i, *i.e.* if i = 5, *pred*(i) returns 4.

For example:

trunc (5.1)	gives	5	*round* (5.6)	gives	6
trunc (−5.1)	gives	−5	*round* (5.1)	gives	5
trunc (5.9)	gives	5	*round* (−5.6)	gives	−6
trunc (−5.9)	gives	−5	*round* (−5.1)	gives	−5

Care should be taken when using *trunc* and *round* because the range of real numbers is far greater than the integer range.

Functions like *round* and *trunc* are called 'type converters' or 'transfer functions' because they provide a method of transferring a value from one type to another. Care must also be taken here not to exceed the integer range when using *trunc* and *round*. Truncation can be used to round-off real values to the nearest integer by:

trunc(x + 0.5)	for x real and x > = 0
trunc(x − 0.5)	for x real and x < 0

Certain functions that are considered as essential by engineers are not given as standard functions *i.e.* tangent, raising a variable to a power, log to the base 10. In

Chapter 5 techniques for writing your own functions to supplement the above list are described.

To evaluate exponentiation use a^n = exp(n*ln(a)).
To evaluate log to the base 10 use log(x) = ln(x)/ln(10).

Worked Example 2.6

Evaluate the following functions and specify the type of result:

    a. *pred*(5)   b. *round* (−9.9)
    c. *trunc* (1.3E19)   d. *ln* (0)
    e. *sqrt* (−64.0)   f. *sin* (45°)
    g. *sqr* (4)   h. *sqr(sqrt*(4))
    i. *succ(succ(succ*(1)))   j. *sqrt* (9) ∗ 3
    k. *succ(pred*(1))   l. *pred* (5) + *round* (5.5)

Solution:

    a. 4 integer   b. −10 integer
    c. semantic error   d. semantic error
    e. semantic error   f. not valid
    g. 16 integer   h. 4.0 real
    i. 4 integer   j. 9.0 real
    k. 1 integer   l. 10 integer

trunc can only handle *real* in fixed point notation.

Input to a Program

An input instruction reads data from the input stream and allocates this value to a variable. In Pascal this is done by the use of two standard identifier procedures (see Table 1.2):

    read   and   readln

One form of a read statement is

    read (variab1,variab2,variab3);

one or more variables may appear inside the parentheses as long as they are separated by a comma. For example:

    read (x,y,beta);
    read (x,y);*read*(beta);   (∗ are all the same ∗)
    read (x);*read*(y);*read*(beta);

Data values are passed to the program from the input device such as a keyboard or a card reader. The association of data value with a variable is done entirely on the basis of position. The first value found in the input stream is associated with the first variable in the *read* list and so on. Data in the input stream are separated by at least one space if more than one data value is supplied on the same line or card. Each variable must have the correct type association *i.e.* an integer value to correspond to an integer variable otherwise an error is flagged (type mismatch). Similarly, *readln* (read line) is used to read in data, but it can also be supplied without parameters. The effect of

    readln;

In some Pascal implementations you can separate your data with commas.

with no parameters is to disregard the remainder of the current line of input. The next data requested is obtained from the next line of data. Using *readln* in the form:

    readln (Res); *readln* (IND);

is particularly useful when it is required only to read part of an input line such as:

    10   Resistors
    56   Inductors

When *readln* is supplied with parameters such as:

    readln (a,b,c);

this is equivalent to

    read (a,b,c,); *readln*;

i.e. the difference between *read* and *readln* is a carriage return. The input data is always read from left to right within each line and from one line to the next. When reading in variables make sure that there is no type mismatch and that data values are assigned to variables of the same type *e.g.* a real data value to a real variable. Once a data value has been read in it can not be re-read (except when using files; see Chapter 6).

Output from a Program

The role of an output instruction is to get results or messages out of the computer in some suitable form *e.g.* printed on paper or displayed on a screen. In Pascal this is done with the standard identifier procedures:

    write   and   *writeln*

Again the only difference between *write* and *writeln* (write line) is a carriage return; *writeln* need not have a parameter.

    writeln;   (* This has the effect of writing a blank line *)

In the simplest form the parameters of *write* or *writeln* may be a number or a string of one or more characters. A sequence of characters enclosed by quotes (apostrophes) forms a 'string'; it is used in a program to write text. If a string contains a quote then the quote must be written twice to distinguish it from the end quote. For example:

> Some compilers do not print an output until the first *writeln* statement is encountered in the program *i.e.* standard (?) Pascal on DEC10.

    writeln('Frequency response');
    writeln('The Op-amp' 's gain is = ');

If a parameter of the *write* procedure is an integer then this is printed out as an integer to a predefined 'field width'. Similarly real variables are printed out in a prescribed format depending on your Pascal implementation.

    The procedure *write* is frequently used to combine string output and data output in one statement *i.e.*

> Volts is a variable previously declared in the program.

    write('The input voltage is = ',Volts);

A very useful combination of *write* and *read* is:

```
write('Enter number of frequency points = ');
readln(freqpts);
```
Make your program as interactive as possible.

Another useful combination, if you have a large amount of output to be listed on a terminal, is:

```
write('Press return to continue');
readln;
```

When real and integer numbers are printed out by a Pascal program they are right justified (right aligned); this is very useful for tabular representation of data.

Formatted Output

If the format field width obtained with a simple *write* statement is not acceptable, then this can be changed within the Pascal program. For example the statement

```
write(integervalue:fw);
```

indicates that the *integer* variable integervalue is printed right justified in a field width specified by the integer number fw. If the integer number requires fewer spaces than fw then an appropriate number of spaces is added to the left of the number. If the specified field width is too small then the specification is ignored and the field width is expanded to the necessary field width.

A useful trick.

Specifying a field width of unity (fw = 1) ensures that the output is printed with the minimum number of spaces.

For real numbers a second format parameter may be specified to indicate the required number of figures after the decimal point. This specification follows the field width specification and is preceded by a colon *i.e.*

dp is an integer.

```
write(realvalue:fw:dp);
```

The real number is rounded to the specified number of decimal places (dp) and is printed in 'fixed point format' within the specified field width. If one format parameter is specified only with *real* data then the output is printed in floating point notation to the specified field width.

For some Pascal implementations page *has no effect.*

Another useful standard Pascal procedure is *page* (see Appendix B) which when called with no parameters will force a page skip if the output is printed on the line printer (form-feed). A few extra lines are printed if the output is produced on a VDU screen. This procedure is useful in separating the output into page sections.

Worked Example 2.7

Write a Pascal program to investigate the field width for printing real and integer numbers using the compiler that is available to you.

Examine each statement carefully and compare with the given output.

```
program FormatTest (output);

(*This program investigates the various output
  formats that are available using Pascal
  and prints maxint.                          *)

const pi=3.141592653589;
var   one,two,five : integer;
```

```pascal
            rl,Null      : real;
begin
  one:=1;two:=22;five:=12345;
  rl:=123.4565656565656;
(* Print a string of number in ascending order to help you count *)
  writeln('Test for integer numbers');
  writeln('123456789012345678901234567890');
  writeln(one);
  writeln(two);
  writeln(five);
  writeln(-five);
  writeln(five:1);
  writeln(five:5);
  writeln(five:7);
(* Text for maximum  integer available in your system *)
  writeln(maxint); writeln(' = maxint');
  writeln;writeln;
(* The above two statements will produce two newlines*)
(* Test for reals now*)
  writeln('Test for real number');
  writeln('123456789012345678901234567890');
  writeln(pi);
  writeln(pi:4);
  writeln(pi:4:2);
  writeln(pi:2:5);
  writeln(rl);
  writeln(rl:6:2);
  writeln(rl:25:1);
  write(10/3*3);writeln(' =(10/3)*3');
  write(0.1*10);writeln(' =0.1*10');
  write(Null);writeln(' = not assigned')
end.
```

The above program was run on a VAX 11/750 supporting Unix and Berkeley Pascal and on a Vector MZ supporting U.C.S.D. Pascal. The output from these two computers is given in Fig. 2.3 side-by-side to ease comparison.

```
OUTPUT FROM PROGRAM :              OUTPUT FROM PROGRAM :
     (UNIX)                            (Vector MZ)

Test for integer numbers           Test for integer numbers
123456789012345678901234567890     123456789012345678901234567890
          1                        1
         22                        22
      12345                        12345
     -12345                        -12345
12345                              12345
12345                              12345
  12345                              12345
2147483647 = maxint                32767 = maxint


Test for real numbers              Test for real numbers
123456789012345678901234567890     123456789012345678901234567890
  3.14159265358900e+00               3.14159
 -3.14159265358900e+00              -3.14159
 3.1e+00                             3.14159
 3.14                                3.14
 3.14159                             3.14159
  1.23456565656566e+02               1.23456E2
 123.57                              123.57
                   123.6                              123.6
 1.00000000000000e+01 = (10/3)*3   1.00000E1 = (10/3)*3
 1.00000000000000e+00 = 0.1*10     1.00000 = 0.1*10
 0.00000000000000e+00 = not assigned  1.15054E-14 = not assigned
```

Fig. 2.3   Output from program Format Test.

As can be seen from Fig. 2.3, the field width and the number of digits given in floating point notation are system-defined. In the Unix system integer values cover a field width of 10 characters and real values cover a field width of 22 characters. In the U.C.S.D. Pascal run on the Vector MZ the field width is left justified, contrary to the standard Pascal specifications. In both systems a shortened real value is rounded to the specified number of decimal points. Of interest is the difference in the two systems for *write*(pi:4). A minor problem with the Unix system is the extra space printed to the left of real values regardless of field width specification.

Notice also that *maxint* in the Vector MZ system is 32767 and in the Unix system is 21474883647.

An important point to note from Fig. 2.3 is that U.C.S.D. Pascal does not automatically set all program variables to zero. A good program should not use non-assigned variables. It is easier to recover from this type of error if the variables are initialised to zero by the programmer. For example, if a non-assigned variable in a division is used this produces a semantic error, whereas with the 1.15054E-14 assigned to identifier Null by the Vector MZ the program continues but produces erroneous results.

Before writing any Pascal program to solve problems make sure that you are aware of the idiosyncrasies of your Pascal. The previous program stresses one of these points. Further tests on Pascal compilers are given in later Chapters.

Using Ohm's Law calculate the current through a resistor, given the voltage across it.

```
program ohmslaw(input, output);
(* This program demonstrates Ohm's law: V = RI
   for a 10 Ohm resistor                         *)

const resistor = 10.0;

var vsource, isource :real;

begin
  write('Enter source voltage: ');
  readln(vsource);
  isource := vsource / resistor;
  writeln('Source voltage = ', vsource);
  writeln('Source current = ', isource);
end.

Execution begins...

Enter source voltage: 12.0
Source voltage  =   1.20000000000000e+01
Source current  =   1.20000000000000e+00

Execution terminated.
```

A Step by Step Development of Simple Pascal Programs

Case Study 1

The simplest bias circuit for a Bipolar Junction Transistor (BJT) is shown in Fig. 2.4. A base resistor R_B provides the appropriate base current for the transistor

Ritchie, G.J. *Transistor Circuit Techniques* (Van Nostrand Reinhold, 1983) pages 32–4.

25

which has a collector load R_C. Calculate R_B and R_C, given that $V_{CC} = 12$ V, $V_{CE} = 6$ V, $I_C = 4$ mA and that the transistor $\beta = 50$.

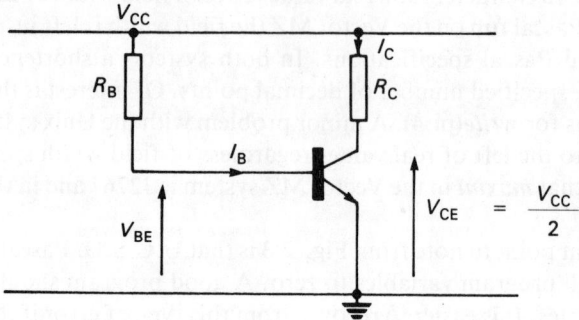

Fig. 2.4 Constant base current bias circuit.

Assuming that a quiescent collector voltage of $0.5 V_{CC}$ is specified and that $V_{BE} = 0.7$ V

$$\text{Quiescent } I_C = \frac{V_{CC}}{2R_C}$$

and

$$\text{Quiescent } I_B = \frac{1}{\beta} \cdot \frac{V_{CC}}{2R_C}$$

The algorithm to calculate R_B and R_C can be expressed in terms of the following equations:

$$R_B = (V_{CC} - V_{BE}) \cdot \frac{\beta}{I_C}$$

and

$$R_C = \frac{V_{CC} - V_{CE}}{I_C}$$

The following Pascal program to obtain a solution to the problem can be written:

```
program BaseCurrentBias1(input,output);
(* This program calculates the base resistor necessary
   to bias a BJT to specified quiescent conditions        *)
const VBE=0.7;VCC=12.0;VCE=6.0;
      IC=0.004;beta=50;
var RB,RC : real;
begin
      RB:=(VCC-VBE)*beta/IC;
      RC:=(VCC-VCE)/IC;
      writeln(RB,RC);
end.
```

Execution begins...

1.41250000000000e+05 1.50000000000000e+03

Execution terminated.

As far as the solution to the problem is concerned the above program gives us the desired result. But the whole essence of writing programs is the ability to use the same program, with the minimum number of modifications, for different cases *i.e.* the previous program is too restrictive. The only way to use this program for other conditions is to use the editor and modify all the constants. This is a very tedious process. A much better approach is to make the program interactive and to enter the various values for the variables at run time. Therefore the program can be rewritten as:

```
program BaseCurrentBias2(input,output);
(* This program calculates the base resistance
   necessary to bias a BJT. It is assumed that
   Vbe=0.7 volts.                                            *)
const Vbe=0.7;
var  Vcc,Vqce,Vce,Iqc,Ic,RB,RC,beta : real;
begin
    write( 'Enter value of dc power supply =');readln(Vcc);
    write( 'Enter quiescent conditions : Vce (volts) and Ic (mA) = ');
    readln(Vqce,Iqc);
    Iqc:=Iqc/1000;(* To convert mA to Amps *)
    write( 'Enter transistor beta = ');readln(beta);
(* Calculations starts from here *)
    RC:=(Vcc-Vqce)/Iqc;
    RB:=(Vcc-Vbe)*beta/Iqc;
(* At this stage we should try to find the nearest preferred values
   but the procedure to do this has been omitted.              *)
    Ic:=beta*(Vcc-Vbe)/RB;
    Vce:=Vcc-Ic*RB;
    writeln;writeln;
    writeln('The required collector resistor RC = ',RC/1000:5:2,'kohms');
    writeln('The required base resistor RB = ',RB/1000:5:2,'kohms');
    writeln('Collector-Emitter voltage = ',Vce:3:1,'volts');
    writeln('Error in collector current = ',(Vqce-Ic)*1000,'mA');
    writeln('Error in Collector-Emitter voltage = ',(Vqce-Vce),'volts')
end.

Execution begins...

Enter value of dc power supply =12
Enter quiescent conditions : Vce (volts) and Ic (mA) = 0.6 1.0
Enter transistor beta = 100


The required collector resistor RC = 11.40kohms
The required base resistor RB = 1130.00kohms
Collector-Emitter voltage = -1118.0volts
Error in collector current =   5.99000000000000e+02mA
Error in Collector-Emitter voltage =   1.11860000000000e+03volts

Execution terminated.
```

Resistor component manufacturers do not produce all possible values but rather they produce a set of preferred values *e.g.* the preferred range of the E12 series is 1.0, 1.2, 1.5, 1.8, 2.2, 2.7, 3.3, 3.9, 4.7, 5.6, 6.8, and 8.2. To construct this circuit preferred values must be used for the components. To do this it is necessary to use boolean conditions to find the nearest preferred values for RB and RC. Also of interest would be the error which is introduced by using preferred values rather than exact values and whether the transistor enters saturation. Chapters 3 and 4 introduce the idea of boolean conditions and control structures.

Ritchie, G.J. *Transistor Circuit Techniques* (Van Nostrand Reinhold, 1983) pages 153–4.

Case Study 2

Electronic filters are a basic component in many communication systems such as telephones, radio, radar, and so on. It is in fact very difficult to think of a moderately complex system which does not contain a filter. The most commonly used one is a Butterworth filter whose magnitude characteristics can be described by the transfer function:

Zverev, A. *Handbook of Filter Synthesis* (Wiley, 1967).

$$\left|\frac{V_o}{V_{in}}(j\omega)\right|^2 = \frac{1}{1+\omega^{2n}} \qquad (2.1)$$

Filter specifications are usually given in terms of attenuation in dB for a specified normalised frequency in the stopband. The frequency response of Butterworth filters for various orders (values of n) are shown in Fig. 2.5.

A short Pascal program can be written to obtain the required order given an attenuation requirement at a specified normalised frequency. Before writing the program a number of points must be considered first. The attenuation is specified in dB therefore it is necessary to have the capability of calculating logarithms to the base ten (log), and also to be able to raise a variable to a given power.

Equation 2.1 can be rewritten as:

$$\left|\frac{V_o}{V_{in}}(j\omega)\right| = 20 \log \frac{1}{(1+\omega^{2n})^{\frac{1}{2}}}$$

$$= -10 \log (1 + \omega^{2n})$$

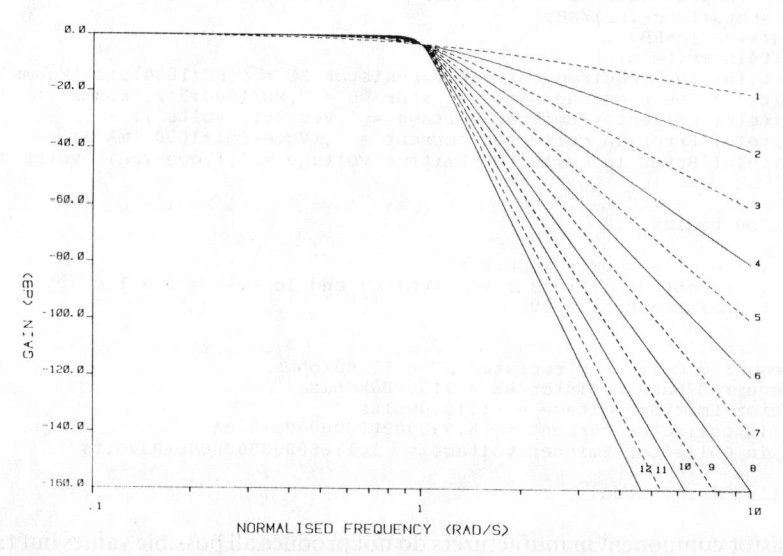

$$= -x \text{ dB} \qquad \text{(say)}$$

Fig. 2.5 Butterworth filters, order 1–12.

From the above equation we can evaluate n as

$$n = \frac{\log (10^{\frac{x}{10}} - 1)}{2 \log \omega}$$

Converting log to logarithm to the base e (ln)

Since log(a) = ln(a)/ln(10).

$$n = \frac{\ln (10^{0.1x} - 1)}{2 \ln \omega}$$

It is also necessary to raise 10 to the power 0.1x. This can be achieved with the use of the identity:

$$a^k = \exp(k * \ln(a))$$

Therefore, in Pascal the order (n) can be calculated using the following assignment statement:

$$n := (ln(exp((x/10)*ln(10)) - 1))/(2*ln(w));$$

The right hand side of the above equation gives a real number and as the required result should be an integer value, the value of n should be evaluated by rounding up the real result. Choosing a higher value of n means that the requirement is surpassed by a certain safety margin. This is done in the given program using the statement:

$$order := trunc(n) + 1;$$

A Pascal program to evaluate the required order for Butterworth filter for a given normalised stopband specification is given here:

```
program ButterworthOrder(input,output);
(* This program calculates the required order of a Butterworth
   filter given the attenuation in dB at a specified normalised
   frequency in the stopband.*)
var dB,w,n,loss : real;
    order       : integer;
begin
  write('Enter attenuation in dB =');readln(dB);
  write('Enter normalised frequency in rad/s =');readln(w);
  dB:=abs(dB);(* just in case it is entered as a negative number *)
  n:=(ln(exp((dB/10)*ln(10))-1))/(2*ln(w));
  order:=trunc(n)+1;
  loss:=10*ln(1+exp(2*order*ln(w)))/ln(10);
  writeln('The required order is =',order:1);
  writeln('The attenuation at ',w:1:2,' rad/s is = ',loss:5:3,' dB')
end.

Execution begins...

Enter attenuation in dB = 60
Enter normalised frequency in rad/s = 10
The required order is =3
The attenuation at 10.00 rad/s is = 60.000 dB

Execution terminated.
```

Summary

A program has basically two types of data that it uses in its execution: constants and variables. The value of a constant does not change throughout the execution of the program whereas variables can be given various values at different stages of the execution. Constants have values of unstructured data types. In contrast, variables can be of structured and/or unstructured data type.

In Pascal any number of identifiers can be defined to represent constant values provided they are declared in the **const** definition part of the program. All variables used in a Pascal program must be declared by specifying their identifier-name and type in the **var** declaration part of the program.

In this chapter two types of simple variable were examined :*real* and *integer*. The range of values that can be assigned to both *real* and *integer* data is restricted by the computer word length. Integers are whole numbers and do not contain a decimal point. Commas are not allowed with any numbers in Pascal; they are solely reserved as separators of items in lists. Real numbers can be expressed in either floating point notation or fixed point notation, but no real number is allowed to begin or end with a decimal point. Either real or integer values can be assigned to *real* variables but only integer values can be assigned to *integer* variables.

Pascal provides arithmetic operators that can be used with *real* and *integer* numbers. To determine the order in which arithmetic operations are carried out, they are given priorities. This assignment of priorities is also extended to other operations, to be met later, such as operational and logical. Parentheses override the usual operator priorities. Use parentheses when you want to dictate the order of execution to the computer rather than leaving the order of execution to the computer.

Pascal also provides two standard identifier procedures to aid the programmer with his input and output. These are *read, readln* and *write, writeln*. The difference between *read* and *readln*, and between *write* and *writeln* is a carriage return. One or more variables can be read into the program by the use of the *read-readln* procedures. The procedures *write-writeln* can output both data and strings. A facility also exists in Pascal to limit the field width of *real* and *integer* data and strings when the *write-writeln* procedures are used.

Problems

2.1 Write valid constant declarations and specify the data type for the following:

(a) 6.28   (b) A   (c) Success
(d) &   (e) FALSE   (f) false
(g) 3 > 2   (h) 365   (i) trunc (1.76)

2.2 Write the Pascal representation for the following expressions:

(a) Area $= \pi r^2$ (b) $y = \dfrac{e^x - e^{-x}}{2}$

(c) $x_1 + x_2 + x_3 = y$ (d) $a = \ln(1 - x^2)^{\frac{1}{2}}$

(e) $a = \sin^2 y + \sin^2 x$ (f) $\omega = \log(\omega_0 + \log\omega_1)$

2.3 Evaluate the valid Pascal expressions indicating the type of result:

(a) 0.07/2.757 * 5 + 0.5 * exp(1)   (b) 3 + 6/3 − 12 div 3
(c) 5 mod 2.5 div 1   (d) trunc(1.7 + 5.3/2) mod 2
(e) 3 + succ(5) * 7/3   (f) pred(5.0)/5
(g) − 77 mod 3 div 3   (h) 5 div trunc(pred(5 mod 2))

2.4 Write the necessary Pascal steps to find the remainder of dividing 32.76 by 3.

2.5 Write Pascal programs to:
  (i) Convert metres to miles.
  (ii) Convert °F to °C.
  (iii) Raise x to the power of y.

2.6 Given the following three lines of input data:

    101.6567   76.645
    − 8
    0.007   0.008   0.009

and the following declarations:

```
a,b,c :real;
i     :integer;
```

What is the effect of executing the following sequence of statements:

(a) readln(a,b);readln(c);write(a:3:2,'next',b:5:2,c);
(b) read(a,b,i);readln;write(i,trunc(a−b):1);
(c) readln(a);readln(b);readln(c);writeln((a+b+c):7:2);
(d) readln;readln;readln(a,b,c);write('The sum is = ',a+b+c:5:3);

2.8 Write a Pascal program to calculate the two port short circuit admittance parameters given the open circuit parameters.
2.9 Write a Pascal program to calculate the first five terms of a cosine series.
2.10 Write a Pascal program to calculate hyperbolic sine and hyperbolic cosine.

3 Scalar Data Type: Char, Boolean, Enumerated and Subrange. The Array Data Structure

Objectives
- ☐ To introduce the scalar data type *char*.
- ☐ To introduce the scalar data type *boolean*.
- ☐ To define the necessary steps for the introduction of user-defined scalar data types: enumerated and subrange.
- ☐ To introduce the basic concepts of an **array**.

In Chapter 2 two of the Pascal-defined scalar data types were introduced: *integer* and *real*. The other two Pascal-defined scalars of type *char* and *boolean* are examined in this chapter. Standard function identifiers that are applicable to these two new types are also introduced.

One of the main advantages of Pascal over many high level languages is the ability of the programmer to create his own type of scalar data type. In this chapter the two user-defined scalar data types are presented; these are the enumerated type and the subrange type. A number of advantages can be obtained from using these user-defined data types.

Arrays are also introduced in this chapter even though they are part of the Pascal structured data types which are examined in detail in Chapter 6. Since even the simplest form of **array** is very useful in solving engineering problems, it is described at this stage. Its introduction here also facilitates the use of program control structures to be examined in Chapter 4.

Computer Character Set

Every computer has a set of symbols or characters such as letters, digits, punctuation marks and control characters by means of which it communicates with its input and output devices. Unfortunately, most computers do not support the same character set. The most common character code is the ASCII (**A**merican **S**tandard **C**ode for **I**nformation **I**nterchange) with a total of 128 printable and control characters. The full ASCII character set is shown in Fig. 3.1.

Some computers use a subset of ASCII

The available character set is defined in the computer according to a predefined sequence (implementation defined) which is represented by consecutive non-negative integer values starting at zero. This sequence is referred to as a 'collating sequence' or a 'lexicographic sequence'. The collating sequence for ASCII characters is shown in Fig. 3.1 where the 'ordinal number' of any character (its position) is given by the sum of its row number and column number.

```
        0   1   2   3   4   5   6   7   8   9  10  11  12  13  14  15
------------------------------------------------------------------------
  0:               ...C O N T R O L
 16:                         C H A R A C T E R S...
 32:        !   "   #   $   %   &   '   (   )   *   +   ,   -   .   /
 48:    0   1   2   3   4   5   6   7   8   9   :   ;   <   =   >   ?
 64:    @   A   B   C   D   E   F   G   H   I   J   K   L   M   N   O
 80:    P   Q   R   S   T   U   V   W   X   Y   Z   [   \   ]   △   -
 96:        a   b   c   d   e   f   g   h   i   j   k   l   m   n   o
112:    p   q   r   s   t   u   v   w   x   y   z   {   :   }   ~
------------------------------------------------------------------------
```

Fig. 3.1 ASCII character set.

Note in ASCII non-printing control characters precede the printable characters. The character space (in position 32) is considered as a printable character; other control characters such as line-feed and carriage return are not.

Most character sets include upper case letters, digits, the space (blank) character, some punctuation marks and some mathematical symbols. You should find out the character set that is available for your use by looking at your Pascal user manual or by writing a simple program to list the character set.

The character set used on CDC 6000 computers has only 64 characters.

The Data Type Character

In Pascal a variable of the data type *char* is defined as a character within the character set available on a computer. Associated with each character is an 'ordinal number' *i.e.* its position within the ordered set. The first character in a set has the ordinal number 0.

In a Pascal program a constant of type *char* is denoted by enclosing a single character between quotes. An exception to this is the quote character itself which is denoted by '' *i.e.* the single quote is written twice. For example:

```
.....................
var initial,star,space,quote:char;
..........
   space:= ' ';
   initial:= 'Y';
   star:= '*';
   quote:= '''';
..........
```

It is important to restate here that char variables are always denoted by a single character and not a string of characters.

An important aspect of the available character set is the order in which the characters are represented in the computer *i.e.* the collating sequence. The ordering varies from computer to computer and it is also Pascal implementation dependent. For example, the letters come before the digits in EBCDIC but the opposite is true in the ASCII case.

For portable programs we can only make the following assumptions:

(1) The digits 0 to 9 are numerically ordered and contiguous.
(2) If the upper case letters A to Z are available then they are alphabetically ordered but not necessary contiguous.
(3) If the lower case letters a to z are available then they are alphabetically ordered but not necessary contiguous.
(4) The ordering relationship between any two character values shall be the same as between their ordinal numbers.

IBM uses the Extended Binary Coded Decimal Interchange Code (EBCDIC) character set with 256 possible characters.

Note that, for example, '9' is not the integer 9 nor is 'A' the identifier A.

For example:

'0' < '1' < ..... < '9'
'A' < 'B' < ..... < 'Z'
'a' < 'b' < ..... < 'z'

If the above assumptions can be accepted then the following comparison or relational operators can be used for comparing characters within a set.

Mathematical symbol	Pascal symbol	Texicographically
=	=	equal to
$\leq$	<=	less than or equal to
<	<	less than
$\geq$	>=	greater than or equal to
>	>	greater than
$\neq$	<>	not equal to

The result from these operations is either *true* or *false*. No other operators such as +, −, sin, and so on, can be used with *char* variables. Relational operators have the lowest priority of all operators after the assignment operator.

Worked Example 3.1

Assume that initial, quote and letter have been declared as *char* variables, can you correct the following statements:

(a) initial = 'K';
(b) letter : 'z';
(c) letter := 'AB';
(d) letter > '9'
(e) initial := 'c' − 'd';
(f) quote = ''';
(g) letter := sqr('1');
(h) 'A' mod 'a';

Solution:

(a) initial := 'K';
(b) letter := 'z';
(c) type mismatch
(d) correct as part of a statement.
(e) not possible
(f) quote := '''';
(g) not possible
(h) not possible

Input and Output of Character Variables

Line-feed is interpreted as a space.

When numerical data is obtained from the input stream using *read*, any preceding spaces or end-of-line characters are ignored. This is not the case when a data of the type *char* is read-in, because a space and a line-feed are considered as valid characters. Assume that chvalue is an identifier of *char* type then

   read(chvalue);

reads-in a single character from the input data stream and assigns it to chvalue, and

   write(chvalue);

writes the character which is stored in chvalue.

There are a few points which should be remembered when dealing with *char* variables:

(i) The character to be read-in by the *read* procedure is not enclosed in quotes in the input data stream.

(ii) Space characters in the input stream are not skipped.

(iii) There are no quotes around chvalue in the *write*(chvalue) statement. If quotes are inserted then the text 'chvalue' is printed and not the character which is associated with the variable chvalue. Similarly if quotes are used with the *read* procedure an error is flagged.

A useful output statement, especially when tabulating results is:

  write(' ':fw);

where fw is the field width specifier *i.e.* if fw = 10 then the above statement produces 10 spaces.

Standard Function Identifiers for Character

There are four standard Pascal functions (see Table 1.2) that are particularly useful for manipulating *char* data. Assuming that ch and int have been declared as *char* and *integer* variables respectively then

ord(ch): The ordinal function: obtains the ordinal number or position of the character ch within the available character set.
  i.e. ord(ch) is 65, if ch was assigned as ch: = 'A'.
      ord('0') is 48

chr(int): The character function: obtains the character whose ordinal number is the non-negative integer int.
  i.e. chr(int) is the character C, if int was assigned as int: = 67.
      chr(57) is the character 9

pred(ch): The predecessor function: yields the previous character to the character ch in the available character set.
  i.e. pred(ch) is the character c, if ch was assigned as ch: = 'd'.
      pred('4') is the character 3

succ(ch): The successor function: yields the next character to the character ch in the available character set.
  i.e. succ(ch), is the character d, if ch was assigned as ch: = 'c'.
      succ('3') is the character 4

Do not confuse the *ord* function with ordinal type.

The ASCII character set is used for these examples.

Although *ord, pred* and *succ* can also be used with integer and boolean variables as arguments, they are rarely used as such. The *ord* and *chr* functions are referred to as 'transfer functions' since they effect a transfer from ordinal numbers to characters and vice versa. The *pred* and *succ* are 'inverse functions' since

  pred(*succ*('A')) is still 'A'
and *succ*(*pred*('A')) is still 'A'

ord and *chr* are also inverse functions.

All standard identifier functions mentioned here give results which are implementation dependent.

Attempts to generate characters outside the given character set causes an error.

Worked Example 3.2 Write a Pascal program to examine the effects of *ord*, *chr*, *pred* and *succ* in your system.

Solution:

```
program  CharacterFunctions(input,output);
var character:char;
    int        :integer;
begin
  write('Enter character = ');readln(character);
  writeln('ord(',character,') =',ord(character));
  writeln('pred(',character,') =',pred(character));
  writeln('succ(',character,') =',succ(character));
  writeln;writeln;
  write('Enter positive integer =');readln(int);
  writeln('chr(',int:1,') =',chr(int))
end.

Execution begins...

Enter character = D
ord(D) = 68
pred(D) = C
succ(D) = E

Enter positive integer = 90
chr(90) = Z

Execution terminated.
```

The above program was run in a number of computers. Some typical results obtained are shown in Fig. 3.2.

Function	U.C.S.D Pascal Vector MZ	Unix Pascal VAX 11/750	Standard Pascal CDC 6000
ord('A')	65	65	1
ord('a')	97	97	not available
ord('6')	54	54	33
ord('#')	35	35	52
ord(' ')	32	32	48
pred('c')	b	b	not available
pred('D')	C	C	C
pred('∧')	]	]	\
succ('Z')	[	[	0
succ('}')	~	~	not available
succ(' ')	!	!	,
chr(52)	4	4	"
chr(67)	C	C	not available
chr(32)	space	space	5
chr(117)	u	u	not available

Fig. 3.2 Output from program character functions.

Note from Fig. 3.2 that *ord*('6') is not the integer 6 but 54 (or 33). This is because the numeral 6 does not occupy the sixth position in any of the character sets used. This problem, however, can be overcome by:

>   integerconversion: = *ord*('6') − *ord*('0');

since all the digits exist numerically ordered and contiguous.

To aid portability of your programs you should always use relative rather than absolute ordinal numbers for the characters, for example

ord('L') — *ord*('A')	is the position of the character L
ord('5') — *ord*('0')	is the position of the character 5
chr(4 + *ord*('0'))	is the character 4

This is sometimes referred to as *boundary condition testing*.

The Data Type Boolean

Boolean variables are declared in the **var** declaration part of the program. Assignment to *boolean* variables can be made with one of only two values: *true* or *false* which are Pascal predefined constant identifiers (see Table 1.2). Similarly boolean constants can only be assigned the values *true* or *false*.

Boolean variables arise from the result of comparisons and are used as flags to control the order by which the statements in a program are executed (see Chapter 4). The relational operators used with *char* variables can also be used to obtain boolean data. The following are valid declarations and assignments for *boolean*:

Boolean are named after George Boole who first used logic on a mathematical basis.

```
          ............
     var  passtest,binary,switchoff:  boolean;
          lowerupper               :  boolean;
          voltage,mark             :  integer;
          ............
          ............
          ............
     passtest   : = (mark > = 50);
     binary     : = true;
     switchoff: = (voltage < 0);
     lowerupper : = ('A' = 'a');
          ............
```

In addition there are three logical operators which can be applied only to boolean constants and variables to produce a boolean result:

>   logical **not** which has greater precedence than
>   logical **and** which has greater precedence than
>   logical **or**

The laws of redundancy, distribution, association, commutation and involution plus De Morgan's theorem are all applicable to boolean variables.

Stonham J. *Digital Logic Technics* (Van Nostrand Reinhold, 1978).

Boolean variables belong to the ordinal type together with integer and character variables. The position of *true* and *false* is predefined in Pascal as:

>   *ord*(*false*) = 0
>   *ord*(*true*) = 1

37

With *write* and *writeln* the word true and false are printed as output for a boolean expression *i.e.*

....
a: = (maxint > 0);
write(a);
....

The word true is printed for a.

Operator Hierarchy

In Chapter 2 the arithmetic operators were discussed and defined their precedence, and in this chapter we have met the relational and boolean operators. It is possible in a complex statement to have both arithmetic and logical operators. In Pascal the hierarchy is defined as

not	which has greater precedence than
* / **div** **mod** **and**	which have greater precedence than
+ − **or**	which have a greater precedence than
= <= < >= > <>	which all have equal precedence.

Most complex statements contain a mixture of arithmetic and logical operators. If the order is ambiguous use parentheses.

If an expression contains more than one operator of the same order of precedence, then evaluation proceeds from left to right, with the overriding rule that subexpressions parentheses are evaluated first.

Standard Functions for Boolean

There are three standard Pascal function identifiers (see Table 1.2) that give results which are of boolean type:

odd(int) : The odd function which has an integer argument. It returns the value true if the argument is odd and false if it is even.

eoln : The end-of-line function with no argument. It examines the next input character without disturbing it in any way and if it is the end-of-line character then the function returns true otherwise it returns false.

A program cannot read the end-of-file character; it can test it only.

eof : The end-of-file function with no argument. It examines the next input character without disturbing it in any way and if it is the end-of-file character then the function returns true otherwise it returns false.

The functions *eoln* and *eof* are examined in greater detail in chapter 6 where file handling is discussed.

Worked Example 3.3

Evaluate the following expressions, assuming that int, ch, test, and value have been declared as *integer, char, boolean* and *real* respectively and initially assigned to:

value: = 9.3;   int: = 11;   test: = true;   ch: = 'y';

(a) test: = (−*maxint* < = int) **and** (int < = *maxint*);
(b) test : = **not** (test);
(c) test: = test **and not** (ch < = *chr*(int ∗ int));
(d) test: = *false* > test;
(e) test: = test **or** *true* **and** test;
(f) test: = ch > *chr*(*ord*(*chr*(99)));
(g) test: = *odd*(value + 1.7);
(h) test: = *odd*(*trunc*(*sqrt*(value)));

Solution:

(a) true and true *i.e.* true
(b) false
(c) true and false *i.e.* false
(d) false
(e) true or true *i.e.* true
(f) true
(g) not possible
(h) true

Scalar Data Type

In Chapter 1 it was stated that one of the most important features of Pascal is the concept of data type and the basic subdivisions were illustrated in Fig. 1.2. The scalar data type subdivision is reproduced here as Fig. 3.3.

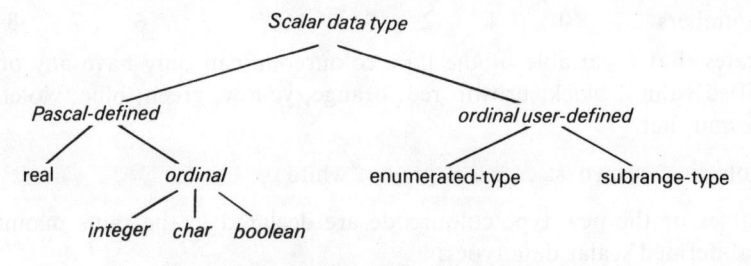

Fig. 3.3 Pascal scalar data type.

The four standard simple or primitive scalar data type identifiers whose properties are defined by Pascal: *integer, char, boolean* and *real* have been discussed. The first three data types are also referred to as 'ordinal' type since they represent an ordered group of values. The total number of constants in the data type is termed the 'cardinality' of the data type. For example, the cardinality of *boolean* is two.

Ordinal means that the values of a given type are ordered and countable and can be compared to each other.

It is possible that none of the four Pascal defined scalar types are suitable for describing a new data type that the programmer wishes to introduce into the program. To facilitate this, Pascal allows two types of user-defined ordinal data types to be defined: (i) the enumerated-type and (ii) the subrange-type.

The enumerated-type defines new types of data that represent scalars (single items) which are interrelated in some numerical order, whereas subrange-type defines data that have contiguous values of an ordinal type and are usually defined as a subset of the ordinal Pascal-defined data type.

A number of programs could be written with only the four Pascal-defined data types but with the use of user-defined types the compiler receives more information about the type of data in the program and thus performs extra checks. Ordinal user-defined types also make the program easier to read and write. By restricting the range of values that can be used with a variable the subrange type minimises possible errors and thus makes the program robust *i.e.* less sensitive to user error or misuse.

Enumerated Scalar Data Type

To create the enumerated data type it is necessary to provide the compiler with the following information: (i) the constant identifiers of the new data type and (ii) their ordering.

The reserved word **type** introduces the new data type in the definition part of the program (see Fig. 1.4). This definition is inserted in the program after the **const** definition and before the **var** declaration. The **type** definition has the format:

> **type**   type-identifier = (const-identifier, const-identifier,....);

The type-identifier is the mnemonic name given to the new data type and const-identifier are the constants of the new data type. The ordering of the constants is determined by the sequence in which the const-identifiers are listed. Enumeration starts from 0. For example:

type colourcode = (black,brown,red,orange,yellow,green,blue,violet,grey,white);
equivalent ordinal
  numbers   0   1   2   3   4   5   6   7   8   9

indicates that a variable of the type colourcode can only have any one of the specified values: black, brown, red, orange, yellow, green, blue, violet, grey or white and that

> black < brown < .... < grey < white

Variables of the new type colourcode are declared in the same manner as the Pascal-defined scalar data types:

> **var**   color:colourcode;

The above definition and declaration can be combined into one:

> **var**   color: (black,brown,red,orange, yellow,green,blue,violet,grey,white);

This short-hand notation is acceptable, but the new data types are not named explicitly as belonging to the type colourcode. Both definitions define the range of colours to which the variable color can be assigned. For example color: = pink will cause an error. Other examples of enumerated data type are:

type units = (pico,micro,milli,kilo,mega,giga);
    weekdays = (Sunday,Monday,Tuesday,Wednesday,Thursday,
                    Friday,Saturday);
    binary = (off,on);

var value:units;
    day:weekdays;
    bits:binary;

The relational operators are applicable to the enumerated scalar types provided that both identifiers are of the same type *i.e.*

Margin notes:

Remember Light Comedy T-V Show (**LCTVS**).

Note the equal sign (=).

You already know the Pascal-defined boolean:
type boolean = (*false,true*);

Note the brackets.

Sunday < giga is not valid whereas micro < milli is

In addition to the six relational operators, we can apply the standard identifier functions: *succ, pred,* and *ord* to the user defined enumerated scalar type. For example using the above definitions:

ord(black) is 0	*succ*(red) is orange	*pred*(green) is yellow
ord(milli) is 2	*succ*(pico) is micro	*pred*(mega) is kilo

Unlike values of the Pascal-defined type, user-defined values cannot be input (*read, readln*) nor output (*write, writeln*) to external devices. Also user-defined scalar type cannot be used to perform arithmetic nor boolean functions *i.e.*

In Chapter 4 the method of overcoming this problem is discussed.

blue + red	Sunday/Saturday	blue **or** green
sqr(brown)	*sin*(pico)	giga **and** micro
read(blue)	*write*(Friday)	*readln*(kilo)

are all illegal.

Subrange Scalar Data Type

The ordinal types *integer, boolean* and *char* represent the entire range of integer, boolean and character values that are available in the system. We can restrict variables to represent part of the range of an ordinal type by defining a new kind of user-defined ordinal type: the subrange. The definition of a subrange specifies the smallest and largest values in the range. Variables of subrange type are declared in the following manner:

type type-identifier = lowerlimit .. upperlimit;
var identifier : type-identifier;

Note only two dots.

The two dots (..) mean through and including. The definition is acceptable if the lowerlimit is less than the upperlimit and that both constant identifiers lowerlimit and upperlimit are of the same ordinal type. For example:

type capletters = 'A' .. 'G';
       numpoints = 1 .. 30;
var AtoG, initial : capletters;
       upto30, timesteps : numpoints;

The short-hand notation can be used with subrange type:

var initial   : 'A' .. 'G';
      numpoints : 1 .. 30;
      AtoG   : 'A' .. 'G';
      upto30   : 1 .. 30;

The use of this short-hand notation is not convenient if there are a number of variables of the same subrange type. The following are typical examples of the use of subrange:

type positive = 1 .. *maxint*;   (* only positive integers *)
       negative = −*maxint* .. −1;   (* only negative integers *)

```
lowfreq   = 0 .. 1000;        (* specify low range *)
midfreq   = 1001 .. 5000;     (* specify mid-range *)
highfreq  = 5001 .. 10000;    (* specifies high range *)
railvoltage = -15 .. 15;      (* confines within a range *)
LtoT      = 'L' .. 'T';       (* forces a limit *)
```

Subrange type scalar data types can also use enumerated-type as the parent type. Parent type is referred to the type from which a subrange can be obtained. For example:

type colourcode = (black,brown,red,orange,yellow,green,blue,violet,grey,white);
　　　lownumbers = black .. green;　(* for the colours representing
　　　　　　　　　　　　　　　　　　　　0, 1, 2, 3, 4, 5　　　　 *)

A number of advantages can be obtained when using subrange:

Use subrange whenever possible.

(1) Programs become more readable and self-documenting.
(2) Validity checks are introduced at run time which add another dimension to error checking.
(3) By reducing the cardinality of a type we reduce the storage space required by the program.

Trying to assign values outside the range of a subrange-type flags an error at run-time because the compiler considers this as an assignment of a different type *i.e.* type mismatch. Subrange type cannot be defined to be of the type *real*.

All operators and functions that are applicable to the parent ordinal type or enumerated type are valid for the subrange type. For example, for the definition

　　type initial = 'A' .. 'R';　　(* subrange of character type *)

then the relational and standard Pascal identifier functions applicable to *char* variables can be used with 'variables' of the type initial. The result produced by such an operation does not necessarily have to lie within the same subrange. The result is considered to be of the parent type, in this case of type *char*.

Some compilers do not allow the result to be outside the specified range.

In contrast with enumerated type, subrange type can be used with the standard procedures *read,readln* and *write,writeln* for input and output as long as the parent type is ordinal Pascal-defined.

Worked Example 3.4　Assume that the following definitions and declarations have been made:

　　type　capletters = 'P' .. 'Y';
　　　　　　spectrum = (yellow,blue,black,violet);
　　　　　　circuit = (current,voltage,resistance);
　　var　　element:circuit;
　　　　　　initial: 'A' .. 'Z';
　　　　　　lastinitial:capletters;
　　　　　　code: (green,orange,red);
　　　　　　colour:spectrum;
　　　　　　test:*boolean*; ch:*char*; int:*integer*;

Evaluate the following expressions:
(a)　int: = *ord*(yellow) + *ord*(resistance) + *ord*('P');　(b)　test: = green < blue;
(c)　lastinitial: = *chr*(*ord*(60) + *ord*(22));　　　　　　(d)　*write*(colour);

(e) resistance: = *trunc*(voltage/current); (f) *write*(yellow > blue);
(d) *readln*(colour); (h) *readln*(lastinitial);

Solution:

(a) 0 + 2 + 80 = 82
(c) R
(e) not possible
(g) not possible
(b) not valid
(d) not possible
(f) false
(h) the first character from the input stream

The Array Data Structure

Arrays are part of the Pascal structured data type (see Fig. 1.2) which vary from the data types already discussed in that they require more than one component for their definition. Four Pascal structured data types: **array, file, record** and **set** are examined in Chapter 6, but the simplest integer and real form of the array data structure is introduced here since it is widely used in this form to solve engineering problems. The **array** structure is re-examined in greater depth in Chapter 6.

An array is defined with three distinct parts:

(i) The array name
(ii) The range of subscripts.
(iii) The type of array components.

The array must be defined in the **type** definition part of the program using the reserved words: **type, array** and **of**

type array-identifier = **array** [subscripts] **of** component-type;

array subscripts must be of ordinal type but its components can be of any Pascal data type.

An array type consists of a predefined number (subscripts) of components which must be of the same type. For example:

```
type  frequencies = array[1 .. 100] of real;   (* array has 100 components *)
      timestep = array[0 .. 30] of integer;    (* array has 31 components *)
      noncausal = array[-10 .. 10] of real;    (* array has 21 components *)
var   omega:frequencies;
      time:timestep;
      rads:noncausal;
```

An alternative symbol for [is (. and for] is .)

Note that the lower limit of an array can be 0 or even a negative value.

The short-hand notation can also be used with **array** *i.e.*

```
var   omega : array[1 .. 100] of real;
      time  : array[0 .. 30] of integer;
```

A one dimensional array can represent a vector.

Note that square brackets are used in the definition and not parentheses. Each component of the **array** can be explicitly accessed by specifying the name of the array and the subscript *i.e.*

omega[1], omega[2], omega[100]

Multidimensional arrays can be declared in the same manner, using a comma to separate the dimensions:

nodalmatrix : **array** [1 .. 20, 1 .. 20] **of** *real*;

The above is a definition of a square matrix with 20 × 20 elements. The elements of this matrix can be obtained using both subscripts *i.e.* the element in the 3rd row and 15th column is contained in 'nodalmatrix[3,15]'.

It is very important to note that the actual size of an array must be declared using constants *i.e.* there are no dynamic array specifications in Pascal *i.e.*

matrix : **array** [1 .. n, 1 .. m] **of** *real*;

must have n and m defined previously as constants.

Worked Example 3.5   Define three array types: (i) a one dimensional array with 7 elements to represent the days of the week, (ii) a two dimensional array with 7∗7 real elements and (iii) a three dimensional array with 7∗7∗7 positive integer elements.

Solution:

    type   weekdays = (Sunday,Monday,Tuesday,Wednesday,Thursday,Friday, Saturday);
             subscript = 0..6;   (∗ if the subscripts start from 0 ∗)
             positive = 1.. *maxint*;
             unitarray = **array** [subscript] **of** weekdays;
             twoarray = **array** [subscript,subcript] **of** *real*;
             threearray = **array** [subscript,subscript,subscript] **of** positive;

Summary

Character processing is highly machine-dependent and there is a need for the programmer to be aware of the character set that is available in his computer. Most computers, except IBM, use the ACSII character set or a subset of it.

In this chapter we have discussed the *boolean* and *char* variables which together with *integer* variables are referred to as ordinal types. An ordinal type is an ordered group of values which are countable and can be compared to each other. Ordinal types and *real* type variables form simple type variables.

A variable of data type *char* is defined as a single character within the available character set. A character in single quotes in a program denotes a constant of type character. But characters in the input stream should not be placed between quotes. A number of standard Pascal functions (*ord, chr, pred, succ*) can be applied to character variables yielding results which are either integer values or other character values. Character variables can be used with the relational operators (=, < =, <, > =, >, < >). When the relational operators compare two characters, and for that matter any two data types, the result of the comparison is of type *boolean*.

Boolean data types are defined to be of ordinal type with their two values defined as *false*<*true*. Boolean types in Pascal may be defined as constants, can be represented by variables, or can be the outcome of a long expression involving relational operators. More complex conditions can be formed by using the boolean operators (**not, and, or**; stated in order of precedence) and the relational operators.

When using relational and boolean expressions, parentheses must be used around the various terms to avoid ambiguity. Boolean operators have a higher priority than relational operators. The primary application of a *boolean* variable is in controlling the execution of a program.

Three standard Pascal function identifiers (*odd, eoln, eof*) produce boolean results when called. The *odd* function has an integer argument and gives *true* if the integer number is odd otherwise the result is *false*. The *eoln* and *eof* functions are very useful when reading data from the input stream; both return *true* if the character read is the end-of-line or the end-of-file character respectively.

The **type** definition was used in this chapter to introduce three new types of data into the program. Two of these are: (i) the enumerated type and (ii) the subrange type. The values included in the **type** definition are given in an ordered fashion. The values of the enumerated type are themselves identifiers but the standard input and output procedures to *read* and *write* cannot be used.

The subrange type represents a contiguous subset of Pascal-defined or user-defined ordinal types. In their definition the smallest and the largest values in the range are specified. Variables of ordinal type should be declared as subrange type whenever possible as this provides a continuous check of the values used for the variables. Two variables do not cause a 'type mismatch' error if they represent values of the same parent type, even though they may be restricted to representing subranges of that parent type.

The **type** definition is also used to define the structured type **array**. An array is considered as a set of subscripted variables. These are distinct variables but all have the same name-identifier. The elements of an array are accessed by location rather than by name. Arrays are very useful in solving engineering type problems.

Problems

3.1 What is the effect of:
    (i)   *chr(ord*(b) −1)   (ii)   *chr(ord*(b) + 1)
    (iii) *ord*('a')−*ord*('A') + 1   (iv) *ord*('6')−*ord*('0')

3.2 What do you understand by:
    (i)   collating sequence   (ii)   cardinality of a data type
    (iii) *char, chr*   (iv) enumerated type
    (v)   subrange type   (vi) **array**
    (vii) ordinal type   (viii) *ord*

3.3 What is wrong with the following definitions and declarations:
    (i)   **type** rainbow = (Red,Orange,Yellow,Green,Blue,Violet);
          var colour,Red: rainbow;
    (ii) **type** realarray = **array** [1 .. 10.0) **of** *real*;
          var A: realarray;
    (iii) **type** body = heart,brain,lung,liver;
          var organ: body;

3.4 Define either an enumerated type or a subrange type for the following:
    (i)   months of the year
    (ii) hours in a day
    (iii) components of a circuit

3.5 Define an array for:

(i) the coefficient of a 10th order polynomial
(ii) the elements of a nodal admittance matrix for a circuit with 6 nodes
(iii) the bits of a byte

3.6 Write a Pascal program to read a character and print its position in the available character set, together with the two previous characters and the two succeeding characters in the set.

3.7 Write a Pascal program to read a date in American form (year/month/day) and output it in British form (day/month/year).

3.8 Assuming that the following definitions and declarations have been made:

```
const  star = '*';
type   positive = 1 .. maxint;
       colour = (red,amber,green);
var    i,j:positive;
       a,b:real;
       test: boolean;
       c1,c2: char;
       lights:colour;
```

State whether or not the following are valid giving your reasons
(a) a:=i+j;
(b) i:=ln(j);
(c) lights:=red+amber;
(d) i:=a **mod** i;
(e) read(colour)
(f) writeln(posinteger);
(g) write(' ':5,star:5);
(h) t = **array**[−1 .. 10] **of** colour;
(i) x = **array**[red..green] **of** positive;

3.9 Given the following 3 input lines:

true   13   1.5
boolean   red
P12absc!/al

what is the effect of the following statements, assuming the declarations given in Exercise 3.8:
(i) read(test,b,a); write(test,b,a:5:3);
(ii) readln; read(c1,c2,c1); write(c1,c2,c1);
(iii) readln;readln;read(c1,c2); write(c2);
    read(c1,c2);write(c1,c2);read(c1,c2);
    read(c1,c2);writeln(c1,c2);

3.10 Evaluate the following expressions, assuming int, ch, test, and value have been declared as *integer, char, boolean* and *real* respectively and initially assigned to:

value := 5.44;   int := 67;   test := true;   ch := 'R';

(i) test := **not**(test **or** test);
(ii) value := value + ord(ch);
(iii) int := ord(test) + ord(ch);
(iv) value := exp(int * ln(2));

Conditional, Repetitive and Goto Statements   4

Objectives

☐ To describe the Pascal statements.
☐ To introduce the Pascal repetitive and conditional structured statements.
☐ To discuss the goto-statement.
☐ To write a program for graphical representation of results.

In Chapter 1 the main subdivisions of a Pascal program were discussed and it was seen that the BLOCK is divided into two parts: (i) the definition and declaration part and (ii) the statement part. Chapters 2 and 3 introduced some of the declarations and definitions which are possible in Pascal. In this chapter the program control statements which are used in Pascal are discussed. This chapter can be outlined very simply using Fig. 4.1

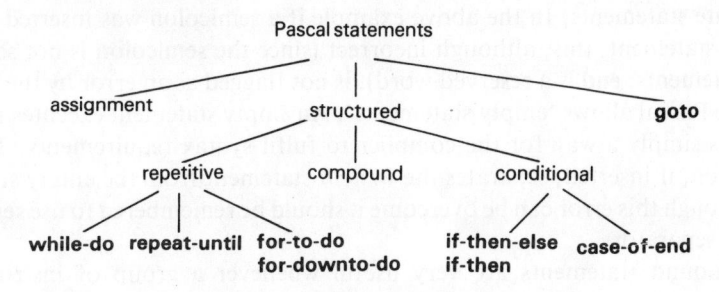

```
                    Pascal statements
           ┌──────────────┼──────────────┐
       assignment      structured        goto
                    ┌──────┼──────┐
                repetitive compound conditional
              ┌────┼────┐         ┌────┴────┐
        while-do repeat-until  if-then-else  case-of-end
                 for-to-do     if-then
                 for-downto-do
```

Fig. 4.1   Pascal statements.

Most of the decision making statements rely on boolean expressions that were encountered in Chapter 3.

Assignment Statement

The assignment statement is the statement most frequently used in a Pascal program and was introduced in Chapter 1. All variables in a program must be initialised before they are used by an expression using an assignment statement or having their initial value read-in. Assignment statements are also used to replace the current value of a variable with the value of an expression. The effect of executing an assignment statement is to evaluate the expression on the right of the 'assignment operator' (:=) and the resultant value is assigned to the variable on the left *i.e.* assignment statements are dynamic within a Pascal program. For example:

    realvalue := 34.75;   (* assignment of a real *)
    charvalue := 'h';   (* assignment of a char *)
    intvalue   := sqr (9) + 5; (* assignment of an integer *)

Whatever value realvalue had previously is discarded and now realvalue is assigned the value 34.75; similarly for charvalue and intvalue.

An error is flagged if both sides of the assignment operator are not of the same type.

Both sides of the assignment operator must be of the same type with the following exemptions:
(i) an integer or integer subrange may be assigned to a variable of type real,
(ii) the type of the expression is a subrange of the type of the variable.

Compound Statement

The compound statement was also discussed in Chapter 1 and it was seen how to treat a number of statements as a single instruction by using the reserved words begin and end as brackets. The statements are executed in the same order as they are written. For example:

It is a good idea to indent a compound statement.

    begin
        write('Enter required order = ');
        readln(order);
        writeln('The order is = ', order)
    end;

Note that the semicolons are used to separate successive statements and not to terminate statements. In the above example if a semicolon was inserted after the *writeln* statement, this, although incorrect (since the semicolon is not separating two statements; **end** is a reserved word), is not flagged as an error by the compiler because Pascal allows 'empty statements'. An empty statement executes no action and it is simply a way for the compiler to fulfil syntax requirements. The extra semicolon, if inserted, separates the *writeln* statement from the empty statement. Even though this error can be overcome it should be remembered to use semicolons only as separators.

Compound statements are very useful whenever a group of instructions is required to be considered as one. Extensive use is made of compound statements in the following sections.

The If Statement

The **if-then-else** structured statement permits the execution of one of two actions by evaluating a specified boolean expression. The form of this statement is:

    if boolean-expression **then** action1 **else** action2;

If the boolean-expression is *true* then action1 is executed and action2 is omitted. If the boolean expression is *false* then action1 is omitted and action2 is executed, *i.e.* one of the two actions are taken depending on the evaluation of the boolean expression. action1 and action2 can be a single statement or a compound statement.

    (* use of single statement *)
    if *odd*(number) **then** *writeln*('The given number is odd')
                else *writeln*('The given number is even');
    ....

```
if x > = 0 then roundreal: = trunc(x + 0.5)
         else roundreal: = trunc(x − 0.5);
```
....
(∗ use of compound statements ∗)
```
if x < = 0 then begin
                    y: = x + 2.67;
                    z: = y + 75 ∗ x
                end
           else begin
                    y: = 0;
                    z: = 0
                end;
```

If action1 or action2 consists of a number of statements make sure that they are given as a compound statement. A lot of people forget!

A very common mistake when using **if-then-else** statements is to insert a semicolon just before the **else**. There is no justification for this.

As it is possible to use compound statements for action1 and/or action2, it is possible to use **if** statements for action1 and/or action2. This case is referred to as a 'nested if' statement.

```
if boolean1 then statement1
            else if boolean2 then if boolean3 then statement 3
                                               else statement 4;
```

Remember no ; before an else.

Great care should be taken when using nested if statement since it is possible to leave variables non-assigned. The rule for unpaired nested **if** statements is that an **else** is always paired with the nearest preceding **then**. Use of proper indentation, as shown above, avoids confusion and makes programs more readable.

The solution to the quadratic equation is a very good example of nested if statements.

```
program QuadraticSolution (input,output);

(* This program calculates the roots of a quadratic equation with
    positive coefficients a,b and c. A number of checks are performed
    for the type of possible solutions.  The imaginary part of a root
    is indicated by j.                                              *)
var a,b,c,re,im,disc :real;
begin
  write('Enter the coefficients a,b and c of a quadratic: ');
  readln(a,b,c);
  if ( a < 0) or ( b < 0) or ( c < 0 )
  then writeln('** Negative coefficients **')
  else
  begin
    if (a = 0) and ( b = 0)
       then writeln('The equation is degenerate')
       else if a = 0
               then writeln('First order equation, root at ',-c/b:1:3)
               else if c=0
                       then begin
                              writeln('Quadratic with c=0, roots at ');
                              writeln(-b/a:1:3,' and 0')
                            end (* with a,b positive and c=0 *)
                       else
                       begin
                         if b = 0
                            then begin
                                   writeln('Quadratic with b=0, roots at ');
                                   writeln('+j',sqrt(c/a):1:3);
                                   writeln('-j',sqrt(c/a):1:3)
                                 end    (* with a,c positive and b=0 *)
```

*It is useful to write a comment after an **end** to describe the previous compound statement.*

Modify this program to allow negative coefficients.

```
                        else begin
                            disc:=sqr(b)-4.0*a*c;
                            re:=-b/(2*a);
                                im:=sqrt(abs(disc))/(2*a);
                                if disc >= 0
                                    then begin
                                        writeln('Real roots at');
                                        writeln(re+im:1:3);
                                        writeln(re-im:1:3)
                                        end (* real roots *)
                                    else begin
                                        writeln('Complex roots at');
                                        writeln(re:1:3,'+j',im:1:3);
                                        writeln(re:1:3,'-j',im:1:3)
                                        end (* complex roots *)
                                end (* with a,c positive and b *)
                        end (* with a,c test for b *)
    end (* positive a,b,c *)
end.

Execution begins...

Enter the coefficients a,b and c of a quadratic: 1 2 1
Real roots at
-1.000
-1.000

Execution terminated.

Execution begins...

Enter the coefficients a,b and c of a quadratic: 1 1 1
Complex roots at
-0.500+j0.866
-0.500-j0.866

Execution terminated.

Execution begins...

Enter the coefficients a,b and c of a quadratic: 1 2 0
Quadratic with c=0, roots at
-2.000 and 0

Execution terminated.

Execution begins...

Enter the coefficients a,b and c of a quadratic: 0 2 1
First order equation, root at -0.500

Execution terminated.

Execution begins...

Enter the coefficients a,b and c of a quadratic: 1 0 1
Quadratic with b=0, roots at
+j1.000
-j1.000

Execution terminated.
```

If it is now required to execute an action only if a certain condition is satisfied without an alternative action, then this can be accomplished by:

if boolean expression **then** action;

The action is not taken if the boolean expression is evaluated as *false*. This is a degenerate case of the **if-then-else** statement with the second option not being available. For example:

```
if value > 0 then write('The number is positive');
if ((value div 2) *2 = value) then write('Even number');
if odd(value) then write('Odd number');
if not odd(value) then write('Even number');
if value > 0 then root: = sqrt(value);
```

Write a Pascal program to evaluate the arctan in either degrees or radians. The program should also distinguish between angles in all four quadrants and should not fail if zero arguments are given.

Worked Example 4.1

Solution:

Most Pascal implementations return the same value for *arctan* in the second and in the fourth quadrant for negative arguments; the result is always given in radians and furthermore the program fails if the argument is zero. The following program evaluates the arctan given two variables x and y. The answer is given in either degrees or radians with an indication of the correct quadrant by defining the angles in the first and second quadrant to lie between 0 and 180° and angles in the fourth and third quadrant to lie between 0 and $-180°$. The possibility of x and/or y being zero also has been taken into consideration.

```
   program atan (input,output);
   (* program eavluates arctan in either degrees or radians.
      Tha angles lie between 0 and 180 or between 0 and -180.*)
   const pi =3.141592653589;
   var phi,efap,x,y,atan:real;
       DegRad  :char;
   begin
     write('Enter value of x and y =');readln(x,y);
     write('Result in D(egrees or R(adians = ');readln(DegRad);
     if x=0.0 then if y > 0.0 then efap:=pi/2
                              else efap:=-pi/2;
     if y=0.0 then if x>=0.0  then efap:=0.0
                              else efap:=pi;

     if (x <> 0.0) and (y <> 0.0) then
        begin
           phi:=arctan(abs(y/x));
           if x > 0.0 then if y > 0.0 then efap:=phi
                                      else efap:=-phi
                      else if y > 0.0 then efap:=pi-phi
                                      else efap:=-pi+phi
        end (* x<>0 and y<>0 *);
     if DegRad = 'D' then begin
                          atan:=efap*180/pi;
                          writeln('Required angle = ',atan:5:2,' degrees')
                          end (* arctan in degrees *)
                     else
                          begin
                          atan:=efap;
                          writeln('Required angle = ',atan:5:2,' radians')
                          end (* arctan in radians *)
end.

Execution begins...

Enter value of x and y = 1 2
Result in D(egrees or R(adians = D
Required angle = 63.43 degrees

Execution terminated.
```

The Case Statement

While with proper indentation the nested if-statement is readable, the logic can become difficult to follow, especially if we have more than three options. Pascal provides the case-statement which selects one statement for execution from a set of statements; the remaining statements are ignored. The selection is based upon the evaluation of an expression of ordinal scalar data type. The case statement is best described with the use of an example:

```
( * assume angle is between 0 and 360 degrees *)
case trunc(angle/90) of
  0: writeln('The angle is in the first quadrant');
  1: writeln('The angle is in the second quadrant');
  2: writeln('The angle is in the third quadrant');
  3: writeln('The angle is in the fourth quadrant')
end (* case angle *);
```

Note the **end** *has no matching* **begin**.

The expression between **case** and **of** is known as the 'selector' or 'case index'. The selector can be of any ordinal data type either Pascal-defined or user-defined. The constants preceding the colon in front of each statement (0, 1, 2, 3) are called 'case labels' which must be of the same ordinal type as the selector and must be unique within the case statement. Also note that the case statement is terminated with an **end**. Since **end** may appear in several contexts in a Pascal program it is advisable to use comments after each **end** as shown in the above example and also in the programs given in the previous section.

The action of the case statement is to evaluate the selector expression and execute only the statement labeled by the resulting value. The statements can be a compound statement or even an empty statement. After execution of the appropriate statement the program proceeds with the execution of the statement following the case **end**. Two or more case labels, separated by commas, can be associated with the same statement. These labels can be written in any order. For example:

In this example the case-selector and the case labels are of type integer.

```
case mark of
  3,2,1,4: writeln('Fail');
  5,6    : writeln('Pass');
  8,7    : writeln('Very Good');
  9,10   : writeln('Credit')
end (* case mark *);
```

The previous examples were based on ordinal type *integer*. It is possible to use the case statement with other ordinal types, for example:

```
    ...................
    (* character type *)
case component of
  'l', 'L':write('Inductor');
  'r', 'R':write('Resistor');
  'C', 'c':write('Capacitor')
end (* case component *);
    ...................
    ...................
```

```
(* enumerated type *)
case units of
  pico  : value:=value*1E-12;
  micro : value:=value*1E-6;
  milli : value:=value*1E-3;
  kilo  : value:=value*1E3;
  mega  : value:=value*1E6;
  giga  : value:=value*1E9
end (* case units *);
```
....

The case statement can be used to output user defined enumerated scalar types such as:

```
case colourcode of
  black  : write('black');
  brown  : write('brown');
  red    : write('red');
  orange : write('orange');
  yellow : write('yellow');
  green  : write('green');
  blue   : write('blue');
  violet : write('violet');
  grey   : write('grey');
  white  : write('white')
end (* case colourcode *);
```

There is an obvious problem with the case statement: if the selector does not take one of the values indicated by the case labels then the case statement is undefined. In standard Pascal a run time error is flagged and program execution terminates. This means that the case statement should only be used if all the possible values which the selector can take are known in advance. Some Pascal implementations ignore the offending case statement and continue with the execution of the program following the case **end**. The best way of avoiding such an error is to ensure that the selector can take only values which exist as case labels. If this cannot be avoided then try printing a sensible error message before program execution terminates. There is no best solution here; it depends on the problem being solved. For example:

Check your Pascal implementation to see how the case statement is implemented.

```
if (value > 0) and (value <= 6))
  then
    case value of
      1: x:=x+1;
      2: begin
           x:=x+2;
           y:=x*x+y
         end;
      3,4,5: ; (* use of an empty statement *)
      6: x:=sqrt(value)
    end (* case value *)
```

Ordering of case labels is arbitrary but they must be unique. An error is flagged if the same case label is used with a number of different statements.

```
                    else writeln(' * * * error case value * * * ');
                    ...................
                    ...................
                    ( * another example * )
                  case basis of
                    'O': base: = 8;
                    'H': base: = 16;
                    'B': base: = 2;
                    'D': base: = 10;
                  others:writeln(' * * error in case basis * * ')
                  end ( * case base * );
```

'Others' case label is used in some versions of Pascal.

The While-Do Statement

Pascal provides two conditional types of repetitive (loop) statements that allow the repetition of other statements to depend upon a boolean condition. These are the while-do and repeat-until statements which are similar in concept and purpose. The while-do statement has the following syntactic form:

while boolean-expression **do** action ;

The action is called the 'component statement' and specifies the action to be repeated. In the majority of cases the component statement takes the form of a compound statement. The boolean-expression is the 'continuation test'. If the boolean-expression is *true* then action is taken otherwise the action is ignored and the program continues execution with the next statement in the program following the while-do statement. Looping continues as long as the boolean expression is evaluated as *true*; it only stops when the boolean expression becomes *false*.

A number of examples of the use of the while-do statement are:

```
                    ...................
                    while (abs(x-a/x) > (x*1E-9)) do x: = 0.5*(x+a/x);
                    ...................
                    ...................
                    ( * factorial of a number * )
                    factorial: = 1;
                    while n > 0 do
                      begin
                        factorial: = factorial * n;
                        n: = n-1
                      end;
                    ...................
                    ...................
                    ( * sum = 1 + 1/2 + 1/3 + .... + 1/n * )
                    sum: = 0; i: = 0;
                    while i < = n do
                      begin
                        i: = i + 1;
                        sum: = sum + 1/i
```

Initial conditions are very important and you should give them a lot of thought.

```
      end;
      ..................
      ..................
   (* find the max and min values in an integer array a[i] (0<=i<=100) *)
      min: = maxint; max: = -maxint; i: = 0;
      while ( (i > = 0) and (i < = 100) ) do
         begin
            if a[i] >max then max: = a[i];
            if a[i] <min then min: = a[i];
            i: = i + 1
         end;
      ..................
      ..................
   (* add all values read-in until end of file is reached *)
      sum: = 0;
      while not eof do
         begin
            read(value);
            sum: = sum + value
         end;
```

*In some implementations, the / can be used only for division of real variables; **div** must be used for division of integers.*

The loop is repeated as long as the boolean expression is true.

A number of important points must be considered here:

(i) The continuation test is evaluated before any action is taken so it is possible that the component statement may never be executed.

(ii) The component statement must contain a statement which eventually alters the value of the boolean expression otherwise the loop executes *forever*.

(iii) Not much attention is paid, by the novice programmer and occasionally by the expert, to the number of times the loop is executed. Usually it is executed once too seldom or once too often. This type of error (called off-by-one) is not easy to detect.

(iv) The initial condition before the test is performed must be carefully considered.

Using user defined identifiers, write a program that grades and sorts examination marks.

Worked Example 4.2

```
program exampressure(input,output);

(* Program grades and sorts exam marks *)

type subjects = (CT105, E100, M100, P110);

var course, bestsubject :subjects;
    mark : array [subjects] of integer;
    grade : array [subjects] of char;
    subnum : integer;

begin
   writeln('Enter marks for CT105, E100, M100 and P110');
   writeln('on separate lines following each prompt');
   writeln;
(* enter subject marks and grades *)
   for course := CT105 to P110 do
   begin
      write('Subject number ', ord(course),': ');
      readln(mark[course]);
```

```pascal
        if mark[course] < 50 then grade[course] := 'N'
        else if mark[course] < 65 then grade[course] := 'C'
        else if mark[course] < 74 then grade[course] := 'B'
        else if mark[course] < 100 then grade[course] := 'A'
    end; (* for course *)

  (* Determine the best subject *)
    bestsubject := CT105;
    for course := E100 to P110 do
      if (mark[course] > mark[bestsubject])
        then bestsubject := course;

  (* print bestsubject *)
    write('Best subject is ');
    case bestsubject of
      CT105: writeln('CT105');
      E100 : writeln('E100');
      M100 : writeln('M100');
      P110 : writeln('P110');
    end; (* case bestsubject *)

  (* request individual grades *)
    writeln;
    writeln('Request for individual grades');
    writeln('Enter subject number (0 - 3) or exit (4): ');
    read(subnum);
    while ((subnum >= 0) and (subnum <= 3)) do
    begin
      course := CT105;
      while ord(course) < subnum do
        course := succ(course);
      case course of
        CT105: writeln('Grade for CT105 is ',grade[CT105]);
        E100: writeln('Grade for E100 is ',grade[E100]);
        M100: writeln('Grade for M100 is ',grade[M100]);
        P110: writeln('Grade for P110 is ',grade[P110]);
      end; (* case course *)
      writeln('Enter subject number (0 - 3) or exit (4): ');
      read(subnum);
    end; (* while *)
end.

Execution begins...

Enter marks for CT105, E100, M100 and P110
on separate lines following each prompt

Subject number          0: 12
Subject number          1: 34
Subject number          2: 56
Subject number          3: 78
Best subject is P110

Request for individual grades
Enter subject number (0 - 3) or exit (4):
0
Grade for CT105 is N
Enter subject number (0 - 3) or exit (4):
1
Grade for E100 is N
Enter subject number (0 - 3) or exit (4):
2
Grade for M100 is C
Enter subject number (0 - 3) or exit (4):
3
Grade for P110 is A
Enter subject number (0 - 3) or exit (4):
4
Execution terminated.
```

Worked Example 4.3 Write a Pascal program to calculate the square root of a positive number using Newton's iterative algorithm:

$$x_{n+1} = \frac{x_n + (number/x_n)}{2}$$

where successive values of x_{n+1} (n = 0, 1, 2, ...) approach the square root of the number. The process is usually terminated when the absolute difference between the given number and the approximation x^2_{n+1} is less than a specified tolerance.

Solution:

Whenever a program of this type is written particular attention must be paid to the value of the specified tolerance and the initial conditions. The initial condition is usually chosen as 1 or as a half the value of the given number (number/2). The tolerance test can be written as:

$$|\, number - x^2 \,| < tolerance$$

but a better and more accurate way to perform this test is

$$|\, number/x^2 - 1 \,| < tolerance$$

which gives sufficient accuracy for very large and very small values of the given number.

```
program SquareRoot (input,output);

(* Program calucate the square root of a given positive
   number.                                              *)

const tolerance = 1E-9;
type counterrange = 0 .. 30;
var number,x :real;
    counter  : counterrange;
begin
 write('Enter positive integer number =');
 readln(number);
 x:=number/2;
 counter:=0;
 while abs(number/sqr(x)-1) > tolerance  do
    begin
      x:=0.5*(x+number/x);
      counter:=counter+1
    end;
 writeln('SQRT(',number:5:3,') = ',x:1:9);
 writeln('Number of iterations = ',counter:1);
 writeln('Pascal sqrt(',number:5:3,') = ',sqrt(number):1:9)
end.

Execution begins...

Enter positive integer number =123
SQRT(123.000) = 11.090536508
Number of iterations = 6
Pascal sqrt(123.000) = 11.090536506

Execution terminated.

Execution begins...

Enter positive integer number =144
SQRT(144.000) = 12.000000000
Number of iterations = 7
Pascal sqrt(144.000) = 12.000000000

Execution terminated.
```

In the above program the *integer* variable counter is used to count the number of iterations that were required to calculate the square root of the given number. By modifying the continuation test of the while-do statement it is possible also to exit from the square root calculation based on the number of iterations (say 10) rather than only on a tolerance value. This can be done by substituting

while ((abs(number/sqr(x)−1) > tolerance) **or** (counter <= 10)) **do**

for the corresponding while-do statement in the program.

The Repeat-Until Statement

Both the repeat-until statement and the while-do statements are used when the exact number of times that the loop needs to be repeated is not known. The repeat-until statement has the syntactic form:

> **repeat** action **until** boolean-expression ;

Compare this definition with the while definition given in the previous section.

The action or 'component statement' consists of one or more Pascal statements which are executed until a boolean expression is satisfied *i.e.* becomes *true*. The statements to be executed need not be formed into a compound statement since the reserved words **repeat** and **until** act as statement brackets.

It should be noted that the continuation test is performed after at least one action has been executed and that one statement within the component statement is required to eventually terminate the loop by altering the value of the boolean expression. For example:

....

The loop is repeated until the exit condition becomes true.

repeat x: = 0.5 ∗ (x + a/x) **until** abs(x−a/x) <= 1E−9;
....
(∗ count characters until a '.' is found ∗)
count: = 0;
repeat
   count: = count + 1;
   read(ch)
until ch = '.';
....

If a semicolon is placed before until one more statement is executed: the empty statement !

....
(∗ count number of digits of an integer number ∗)
digits: = 0;
repeat
   number: = number **div** 10;
   digits: = digits + 1
until number = 0;
....
....
(∗ multiply 10 numbers ∗)
count: = 0; product: = 1;
repeat
   read(number);
   product: = product ∗ number;

Take care with initial conditions.

```
        count: = count + 1
    until count = 10;
        ........................
        ........................
    (* avoid case errors *)
    repeat
        write('test character ? = ');read(ch)                                A useful test.
    until ( (ch = 'A') or (ch = 'B') or (ch = 'C') );
    case ch of
        'B':x: = x*x;
        'C':x: = sqr(x);
        'A':x: = sin(x)
    end (* case ch *);
```

The while-do and repeat-until statements differ in the following ways:

(i) The continuation test for the while-do statement is performed before any action is taken, whereas for the repeat-until statement the test is carried out after at least one action has been taken.

(ii) The component statements in the while-do statement must be formed into a compound statement; this is not necessary for the repeat-until statement.

(iii) The while-do statement terminates when the boolean condition is *false*; the converse is true for the repeat-until statement.

(iv) It is possible with the while-do statement not to take any action; with the repeat-until it is necessary to execute the component statements at least once.

For both while-do and repeat-until statements it is necessary to ensure that within the loop there is a statement which can alter the value of the boolean exit condition.

Write a Pascal program to convert a given positive decimal integer number to an octal, binary or hexadecimal number.   **Worked Example 4.4**

Solution:

```
program Conversion(input,output);
(* Program converts a given positive decimal number to an
   octal or hexadecimal or binary equivalent              *)
type nonegative = 0 .. maxint;
var  decimal,maxpower,base,basepower,digit: nonegative;
     basis,again                          : char;
begin
 repeat (* calculations are not repeated if again='E' or 'e' *)
    repeat (* ensures that the number is positive *)
        write ('Enter positive integer number = ');
        readln(decimal)
    until decimal > 0;
    repeat (* ensures that only 'B' or 'O' or 'H' are accepted *)
      write('Enter required base O(ctal,H(exadecimal or B(inary = ');
      readln(basis)
    until ((basis='O') or (basis='H') or (basis='B'));
    case basis of
        'B':base:=2;
        'O':base:=8;
        'H':base:=16
    end (* case basis *);
    write('Required number = ');
    (* calculate max power of new base *)
    maxpower:=decimal div base;
```

```
              basepower:=1;
              repeat
                basepower:=basepower*base
              until basepower > maxpower;
              (* calculate individual digits of new base *)
              repeat
                 digit:=decimal div basepower;
                 if ((basis='H') and (digit > 9)) then
                    case digit of
                       10:write('A');
                       11:write('B');
                       12:write('C');
                       13:write('D');
                       14:write('E');
                       15:write('F');
                    end (* case ditit *)
                 else write(digit:1);
                 decimal:=decimal mod basepower;
                 basepower:=basepower div base
              until basepower=0;
            writeln(' ',basis);
            write('Type any character to restart or E to exit =');
            readln(again);writeln
         until ((again='E') or (again='e'))
      end.

      Execution begins...

      Enter positive integer number = 56
      Enter required base O(ctal,H(exadecimal or B(inary = H
      Required number = 38 H
      Type any character to restart or E to exit =E

      Execution terminated.
```

Worked Example 4.5 Write a Pascal program that analyses the diode-resistor circuit shown. See G.J. Ritchie, *Transistor Circuit Techniques: discrete and integrated* (2nd edition), Van Nostrand Reinhold (Int.) 1987, Chapter 1.

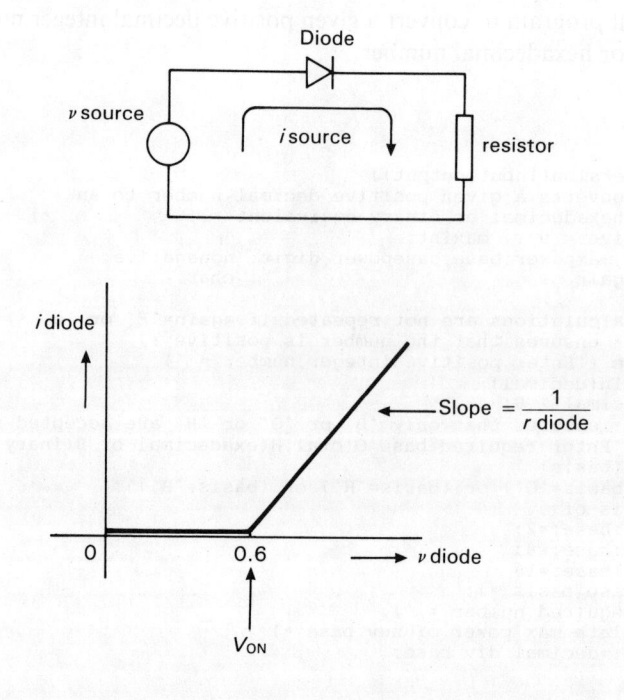

```
program diode(input,output);

(* This program determines the response of a
   diode-resistor circuit to a sinusoidal input *)

const von = 0.6; rdiode = 10.0; pi = 3.14159265;

var vsource, isource, resistor : real;
    time, timestep, amplitude : real;
    step, nsteps : integer;
    turnon : boolean;

begin
  writeln('This program computes the current in a');
  writeln('simple diode-resistor circuit');
  writeln;
  write('Enter resistor value: '); readln(resistor);
  write('Enter amplitude of sinusoid: '); readln(amplitude);
  write('Enter number of input samples: '); readln(nsteps);
  timestep := 2.0 * pi / (nsteps -1);
  time := -timestep;
  writeln('_Voltage_____Current');
  for step := 1 to nsteps do
  begin
    time := time + timestep;
    vsource := amplitude * sin(time);
    turnon := (vsource > von);
    if turnon then
      isource := (vsource - von)/(resistor + rdiode)
    else
      isource := 0.0;
    writeln(vsource:10:4,isource:10:4)
  end
end.

Execution begins...
This program computes the current in a
simple diode-resistor circuit

Enter resistor value: 1000
Enter amplitude of sinusoid: 5
Enter number of input samples: 8
_Voltage_____Current
    0.0000    0.0000
    3.9092    0.0033
    4.8746    0.0042
    2.1694    0.0016
   -2.1694    0.0000
   -4.8746    0.0000
   -3.9092    0.0000
   -0.0000    0.0000
Execution terminated.
```

The For-Statement

The for-statements are used when the number of loop repetitions is known or can be calculated in advance. In this case the number of repetitions does not depend on the effect of statements within the loop but on a variable which is incremented or decremented by one. The variable is given an initial and final value for the loop. There are two forms of for-statement: an incremental and a decremental form. The Pascal syntax of these is:

for identifier: = expression1 **to** expression2 **do** action; (* incremental *)

for identifier: = expression1 **downto** expression2 **do** action; (* decremental *)

The identifier is also called the 'control variable' and must be of ordinal data type *i.e. real* variables are not allowed. expression1 and expression2 must be of the same ordinal data type as the control variable. A for-statement will never be executed if expression1 > expression2 in the case of **for-to** and if expression1 < expression2 in the case of **for-downto**. When the for-statement is first entered the control variable is assigned an initial value given by expression1 and an action is taken. The control variable is then incremented/decremented and action is taken again. The procedure is repeated until the control variable is assigned the value given by expression2, the loop then is executed for the last time, and program execution continues with the statement following the for-statement. action can be any Pascal statement *e.g.* assignment, compound and so on. For example:

```
(* sum = 1+1/2+1/3+ ... +1/10 *)
sum: = 0;
for i: = 1 to 10 do
    sum: = sum + 1/i;
.........................
for character: = 'A' to 'R' do
    write(character);
for character: = 'p' downto 'a' do
    write(character);
.............................
.....................
(* find the max and min value of an array a[i] (0 < = i < = 100) *)
min: = maxint; max: = -maxint;
for i: = 0 to 100 do
    begin
        if a[i] > max then max: = a[i];
        if a[i] < min then min: = a[i];
    end;
.....................
(* even numbers only *)
x: = 0;
for j: = 22 downto 2 do
    if ( j mod 2 = 0 ) then x: = x + exp(ln(2)*j);
(* odd numbers only *)
x: = 0;
for k: = 1 to 11 do
    if ( odd(k)) then x: = x-exp(ln(3)*k);
.....................
```

Note no semicolon after do, otherwise you will be executing an empty statement.

Always check the initial conditions before you enter a loop.

The values that specify the limits could be written as expressions.

The control variable must be of an ordinal type to which the standard Pascal identifiers *succ* and *pred* can be applied. An incremental for-statement of the form:

for i: = initial **to** final **do** action

is equivalent to

```
if initial < = final then
begin
    i: = initial; action;
    i: = succ(i);action;
```

>   ....
>   ....
>   i: = final;action
>   end;
>   (∗ at this point, the i's value is undefined ∗)

and a decremental for-statement of the form:

>   **for** i: = initial **downto** final **do** action;

is equivalent to

>   **if** initial > = final **then**
>   **begin**
>   i: = initial; action;
>   i: = *pred*(i);action;
>   ....
>   ....
>   i: = final;action
>   **end**;
>   (∗ at this point, i is undefined ∗)

As mentioned above it is possible to use any Pascal statement to be repeated by the for-statement. If another for-statement is used, this is referred to as a 'nested' (compound) for-statement. This type of statement is very useful when dealing with matrices or determinates. For example:

>   **for** i: = 1 **to** 10 **do**
>   **for** j: = 1 **to** 10 **do**
>   A[i,j]: = 1;

The above example sets all the components of the two dimensional (matrix) variable A[i, j] to unity a row at a time. This is done by first assigning 1 to i and stepping j from 1 to 10, then i: = 2 and again stepping j from 1 to 10 and so on until i: = 10 and j: = 10 when the final assignment (A[10, 10]: = 1) is done.

A number of points should be stressed at this stage:
   (i) The control variable and the expressions used to calculate the initial and final values must be of the same ordinal data type.
   (ii) The value of the control variable is incremented/decremented automatically every time the loop is repeated until the control variable reaches the final value.
   (iii) The initial and final values for the control variable are evaluated once at the start of the for-statement *i.e.* the number of iterations cannot be modified from within the loop.
   (iv) Incrementing or decrementing is done in steps of one.
   (v) In the incremental for-statement expression1 must be less than expression2 otherwise no action is taken. The converse is true for the decremental for-statement.
   (vi) The value of the control variable is undefined upon exit from the for-statement *i.e.* it must be re-assigned before it can be used again.
   (vii) The control variable can be used within the loop but cannot be changed there. Attempts to assign a new value to the control variable within the loop causes an error.

Many of you will violate (vii)

The newly accepted Pascal standard is reprinted in Wilson, I.R. and Addyman, A.M. *A Practical Introduction to Pascal-with BS6192* (MacMillan, 1982).

(viii) The newly-accepted Pascal standard imposes some restrictions on the control variable which are relevant to **procedure** and **function**. These are discussed in the next chapter.

Worked Example 4.6 Write a Pascal program that uses text characters to draw figures on your computer screen given the x – y coordinates.

```
program display(input,output);

(* Program uses text characters and a 'frame buffer'
   to update a graphics screen                       *)

const xscale = 60; yscale = 20;
      blank = ' '; astrk = '*';

var i,j,n,steps : integer;
    x,y,x1,y1,x2,y2,dx,dy,dxstep,dystep : real;
    display : array [0..xscale, 0..yscale] of char;

begin
  (* Initialize the frame buffer *)
  for j := 0 to yscale do
    for i := 0 to xscale do
      display[i,j] := blank;

  (* Display a blank screen *)
  for j := 0 to yscale do
  begin
    for i := 0 to xscale do
      write(display[i,j]);
    writeln
  end;
  writeln;writeln;
  (* Enter line coordinates *)
  write('Enter (x1, y1): ');
  readln(x1,y1);
  write('Enter (x2, y2): ');
  readln(x2,y2);
  repeat
    x1 := x1 * xscale; y1 := y1 * yscale;
    x2 := x2 * xscale; y2 := y2 * yscale;
    dx := x2 - x1;
    dy := y2 - y1;
    if (abs(dy) > abs(dx)) then
      steps := round(abs(dy)+1)
    else
      steps := round(abs(dx)+1);
    dxstep := dx / steps;
    dystep := dy / steps;
    x := x1;
    y := y1;
    for n := 1 to steps+1 do
    begin
      display[round(x),round(y)] := astrk;
      x := x + dxstep;
      y := y + dystep;
    end;
    (* display the frame buffer *)
    for j := 0 to yscale do
    begin
      for i := 0 to xscale do
        write(display[i,j]);
      writeln
    end;
    writeln;writeln;
    write('Enter (x1, y1): ');
    readln(x1,y1);
    write('Enter (x2, y2): ');
    readln(x2,y2);
```

```
    until ((x1 = x2) and (y1 = y2))
end.

Execution begins...


Enter (x1, y1): .1 .1
Enter (x2, y2): .9 .9

        **
         ***
            ***
               ***
                  ***
                     ***
                        ***
                           ***
                              ***
                                 ***
                                    ***
                                       ***
                                          ***
                                             ***
                                                ***
                                                   ***
                                                      **

Enter (x1, y1): .7 .2
Enter (x2, y2): .3 .8

        **
         ***                                        **
            ***                                  ***
               ***                            ***
                  ***                      ***
                     ***                ***
                        ***          ***
                           ***     ***
                              ** ***
                               ***
                              ** ***
                           ***     ***
                        ***          ***
                     ***                ***
                  ***                      ***
               ***                            ***
            ***                                  ***
         ***                                        **


Enter (x1, y1): 0 0
Enter (x2, y2): 0 0

Execution terminated.
```

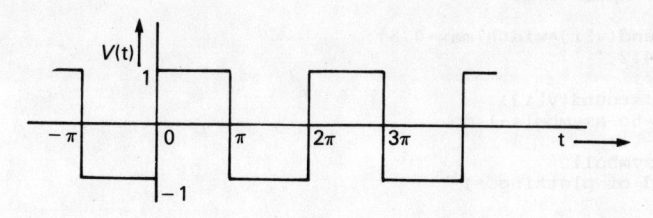

Fig. 4.2 Periodic square wave.

Worked Example 4.7

Fidler, J.K. *Introductory Circuit Theory* (McGraw Hill, 1980).

Write a Pascal program to approximate the periodic square wave shown in Fig. 4.2 using Fourier series. Give your results in a simple graphical form.

Solution:

The approximation to the given periodic square wave can be obtained, noting the symmetry, as:

$$V(t) = \frac{4}{\pi}\left[\sin t + \frac{1}{3}\sin 3t + \frac{1}{5}\sin 5t + \ldots + \frac{1}{n}\sin nt + \ldots\right]$$

where n is odd.

```
program FourierApprox(input,output);

(* This program generates graphical fourier approximations *)

const pi=3.1459265359;  width=40;  symbol='*';
type nonegative = 0 .. maxint;
var t,step,sinesum,max,min:real;
    i,j,k,npoints,nsymbols,nharmonics: nonegative;
    v:array [0..100] of real;
begin
 write('Enter time step and number of points = ');
 readln(step,npoints);
 write('How many harmonics ? = ');readln(nharmonics);
 writeln;writeln;
 (* Calculate the Fourier series *)
 t:=0;
 for i:=0 to npoints do
  begin
   k:=1;  sinesum:=0;
   (* Calculate the odd sine terms only *)
   for j:=1 to nharmonics do
    begin
     sinesum:=sinesum+1/k*(sin(t*k));
     k:=k+2
    end (* j loop *);
   v[i]:=(4/pi)*sinesum;
   t:=t+step
  end (* i loop *);
(* Find the max and min values of the function *)
max:=v[0];   min:=max;
for i:=0 to npoints do
 begin
  if v[i]>max then max:=v[i];
  if v[i]<min then min:=v[i]
 end (* i loop *);
writeln(' ':5,'Approximation of a square wave using Fourier series');
writeln(' ':10,'(4/pi)*(sint + ... + 1/',k-2:1,'*(sin',k-2:1,'t))');
writeln(' ':12,'miny = ',min:4:2,' ':4,'maxy = ',max:4:2);
writeln;
(* prepare data for plotting min to max *)
for i:=0 to npoints do
 v[i]:=v[i]-min;
max:=max-min;
(* plot data *)
t:=0;
for i:=0 to npoints do
 begin
  v[i]:=round(v[i]*width/max+0.5);
  write(t:4:2,'|');
  t:=t+step;
  nsymbols:=round(v[i]);
  for j:=1 to nsymbols-1 do
   write('-');
  writeln(symbol)
 end (* end of plotting *)
end.
```

Note the difference from the previous examples in the calculation of max and min value.

Typical input and output for program FourierApprox is shown in Fig. 4.3

```
Execution begins...

Enter time step and number of points = .2  20
How many harmonics ? = 3

        Approximation of a square wave using Fourier series
             (4/pi)*(sint + ... + 1/5*(sin5t))
                 miny = -1.17     maxy = 1.19
0.00!-------------------*
0.20!------------------------------*
0.40!----------------------------------*
0.60!------------------------------------*
0.80!-----------------------------------*
1.00!---------------------------------*
1.20!---------------------------------*
1.40!-----------------------------------*
1.60!-------------------------------------*
1.80!--------------------------------------*
2.00!-------------------------------------*
2.20!-----------------------------------*
2.40!---------------------------------*
2.60!----------------------------------------*
2.80!-------------------------------------*
3.00!-------------------------*
3.20!----------------*
3.40!-----*
3.60!*
3.80!*
4.00!---*

Execution terminated.
```

Fig. 4.3   Typical input and output for Fourier Approx.

The Goto Statement

The goto-statement is a simple statement indicating that program execution should continue at another part of the program. This transfer of execution is called a *jump* or a *branch* and is accomplished by the statement:

   goto statement-label;

where statement-label is a label that exists in both the statement part of the program and the **label** declaration part of the program block (see Fig. 1.4). All labels used in a program must be declared. This is done before the **var** declaration part with the use of the reserved word **label:**

   program TestLabel(*input,output*);
   label 1,33,4444;
   ....
   ....

L C T V S

In the above example three different labels, separated by comma, have been declared. A label can be any unsigned integer number up to four decimal digits long. The statement where execution is transferred to is prefixed by the label:

   goto 33;
   ....
   ....
   33: statement;

Up to 10,000 labels can be used.

A number of examples are:

```
            read(character);
            if ( character = eof ) then goto 1111;
               ..............
               ..............
      1111:writeln(' * * premature ending of data * * ')
            end;
               ..............
            if divider = 0 then goto 999;
               ..............
               ..............
       999:writeln(' * * Division by zero if continued * * ');
            end;
```

Do not confuse case-labels with labels.

A number of points to remember are:
(i) Labels must be unsigned integers in the range 0 to 9999.
(ii) Labels must be declared only once within a program.
(iii) If a label is declared it must be used.
(iv) A label must be used to prefix one statement only, otherwise ambiguity arises.
(v) A label and its colon form a statement, even if no action follows the colon *e.g.* 99:**end**.

You cannot jump into an if-statement.

(vi) It is not valid to jump with a goto-statement into a conditional or repetitive statement.

A goto-statement can direct program execution either forward or backwards within a program; all statements between the **goto** and the labeled statement are ignored. It is possible to synthesise all the control statements discussed in this chapter using only **if-then** and **goto** statements. This is counter-productive since by doing so we destroy the structured nature of Pascal. Many computer scientists dislike the goto-statement and actively advise against its use. However, there are situations in which the program structure is made clearer if we use goto-statements. For example:

(i) The input data is incorrect or you have reached the end-of-file unexpectedly; in this case the program should write a warning message and exit from the program as quickly as possible.
(ii) To simplify nested logic which sometimes can become very cumbersome.
(iii) As a way to exit from the middle of a loop rather than from the top (while-do statement) or from the end (repeat-until statement).

It has been argued that all **goto** statements can be substituted by careful arrangement of conditional statements.

Inspite of the above it is still advisable to use the goto-statement sparingly. Try first the other control structures and if it is not possible to write the program without goto-statements then, and only then, use them. Goto-statements should only be employed to move forward in a program; if it is necessary to move backwards it means insufficient thought has been given to the problem and most certainly a goto-statement is not required. Try to think in Pascal not in BASIC or FORTRAN. Excessive use of goto-statements results in a program which is difficult to follow and understand.

Worked Example 4.6 Use a goto statement to exit from a program if any element of an integer array is zero.

Solution:

```
program GotoTest(input,output);

(* Tests goto statements *)

label 11;
const n=10;
type posinteger = 1 .. maxint;
var i,j: posinteger;
    number:integer;
    count:real;
    A:array [1..n,1..n] of posinteger;
begin
 count:=0;
 for i:=1 to n do
  for j:=1 to n do
   begin
     read(number);
     if number <> 0 then begin
                          A[i,j]:=number;
                          count:=count+1
                         end
                    else begin
                          writeln(' ** zero detected **');
                          writeln('Position of error = ',count:1);
                          goto 11 (* stop execution immediately *)
                         end
   end;
11:
end.

Execution begins...

12 34 65 67 0

 ** zero detected **

Position of error = 4.0e+00

Execution terminated.
```

Case Study 1

As mentioned in Chapter 2 it is necessary to know the dynamic range that *real* data can be represented by the computer in use (see Fig. 2.2). This can be done by writing a simple program and letting it run until a floating point error terminates execution. There are usually two errors to consider; a 'floating point overflow' error and a 'floating point underflow' error. A floating point overflow error is flagged whenever the result of an expression is greater than the largest possible value by which a real can be represented. A floating point underflow is caused if the evaluation of an expression results in a number which is smaller than the smallest available real number in the computer.

The programs to find these limits are listed below:

```
program Overflow(output);              program Underflow(output);

(* tests for the overflow condition *) (* tests for the underflow condition *)

const forever=50;                      const forever=50;
var i:integer;                         var i:integer;
    x:real;                                y:real;
```

69

```
begin                              begin
  x:=1.1E+20;                        y:=1.1E-20;
  for i:=1 to forever do             for i:=1 to forever do
    begin                              begin
      x:=10*x;                           y:=0.1*y;
      writeln('x =',x)                   writeln('y =',y)
    end                                end
end.                               end.

Execution begins...                Execution begins...

x = 1.10000000000000e+21           y = 1.10000000000000e-21
x = 1.10000000000000e+22           y = 1.10000000000000e-22
x = 1.10000000000000e+23           y = 1.10000000000000e-23
x = 1.10000000000000e+24           y = 1.10000000000000e-24
x = 1.10000000000000e+25           y = 1.10000000000000e-25
x = 1.10000000000000e+26           y = 1.10000000000000e-26
x = 1.10000000000000e+27           y = 1.10000000000000e-27
x = 1.10000000000000e+28           y = 1.10000000000000e-28
x = 1.10000000000000e+29           y = 1.10000000000000e-29
x = 1.10000000000000e+30           y = 1.10000000000000e-30
x = 1.10000000000000e+31           y = 1.10000000000000e-31
x = 1.10000000000000e+32           y = 1.10000000000000e-32
x = 1.10000000000000e+33           y = 1.10000000000000e-33
x = 1.10000000000000e+34           y = 1.10000000000000e-34
x = 1.10000000000000e+35           y = 1.10000000000000e-35
x = 1.10000000000000e+36           y = 1.10000000000000e-36
x = 1.10000000000000e+37           y = 1.10000000000000e-37
x = 1.10000000000000e+38           y = 1.10000000000000e-38

"Overflow"+4 near line 12.         "Underflow"+4 near line 12.

Execution terminated abnormally.   Execution terminated abnormally.
```

For a VAX 11/750.

Case Study 2

For many engineering problems a graphical display of the results is very useful when trying to learn new concepts. In Worked Example 4.7 a graphical representation of the results rather than a tabulation was given, since it is easier to appreciate that a better approximation to the periodic square wave can be obtained by increasing the number of terms in the sine series. This can be done very easily by running the program and increasing the number of harmonics to see the effects of the approximation.

In this section a program is discussed which is capable of plotting the response of up to 9 different curves on the same graph. Furthermore, both the x-axis and the y-axis have numeric scales and a grid is superimposed on the graph. First consider the grid which shown in Fig. 4.4. Ideally it is required to expand the number of x and y points to as many plot units as possible, but a number of physical limitations must be considered. If it is required to display the graph on a VDU then the x units should be limited to 60, and for a line printer to about 110. The y units can be as many as desired, but for a full VDU screen the limit should be set to 40. The limits for the grid shown in Fig. 4.4 are maxxscale = 60 and maxyscale = 20.

The basic idea behind this plotting program is to generate an **array** (called a line) which contains all the x points for each value of y. This array contains all the different characters which are plotted by each line.

Each curve is plotted using a different symbol. The character 1 is used for the first curve, the character 2 for the second and so on. The graph limits are set by finding the maximum and minimum values of x and y from all the curves to be plotted.

```
+---------+---------+---------+---------+---------+---------+
:         :         :         :         :         :         :
:         :         :         :         :         :         :
:         :         :         :         :         :         :
+---------+---------+---------+---------+---------+---------+
:         :         :         :         :         :         :
:         :         :         :         :         :         :
:         :         :         :         :         :         :
:         :         :         :         :         :         :
+---------+---------+---------+---------+---------+---------+
```

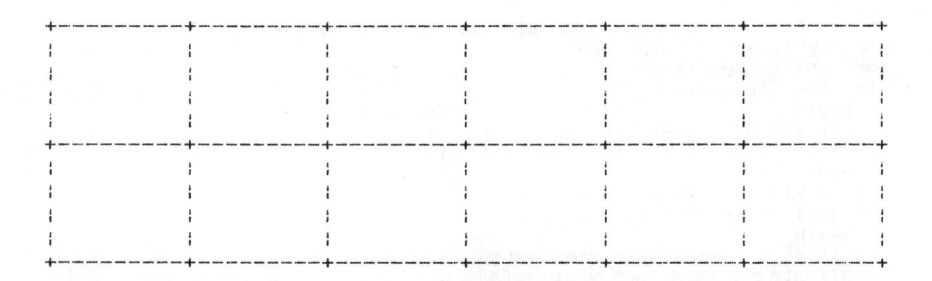

Fig. 4.4 Coordinate axes grid for graph.

The program is self-explanatory and is given below. The following key identifiers help in the understanding of the program:

y : a two dimensional array indicating the curve and its y values.
x : a one dimensional array containing the x values of a particular curve.
line : a one dimensional array containing both the grid symbols and the various curve symbols.

In the given program three test results are also given:

$$\text{Curve 1} = 50*\exp(x)$$
$$\text{Curve 2} = 2*\exp(-x)$$
$$\text{Curve 3} = 30.0$$

The output from this program is given in Fig. 4.5.

```
program Graph(output);

(* Program draws co-ordinate axes for graphs.Up to 9
   plots can be plotted                              *)

const maxxscale=61;
      maxyscale=41;
type posinteger = 0 .. maxint;
var ymax,xmax,xmin,ymin,yrange,xrange,yscale,variable:real;
    nplots,npoints,ysepar,ypoints,yaxis,xpoints,yvalue,xvalue: posinteger;
    i,j,k,m: posinteger;
    blank,bar:char;
    y : array [1..9,1..100] of real;
    x : array [1..100] of real;
    symbol : array [1..9] of char;
    line : array [1..100] of char;
    xscale : array [1..8] of real;
 begin
  (* Test results *)
  nplots:=3; npoints:=40;
  variable:=0;
  for m:=1 to npoints do
   begin
     variable:=variable+0.1;
     x[m]:=variable;
     y[1,m]:=50*exp(-variable);
     y[2,m]:=2*exp(variable);
     y[3,m]:=30.0
   end;
  (* end of test results *)
  for i:=1 to nplots do
   symbol[i]:=chr(i+ord('0'));
```

The letter A, B,... could be used instead of numbers for symbols.

```pascal
                        (* *** Find min and max from all abailable data *** *)
                        ymax:=y[1,1];    ymin:=ymax;
                        for i:=1 to nplots do
                         for j:=1 to npoints do
                            begin
                              if y[i,j] > ymax then ymax:=y[i,j];
Scan for max y-value.         if y[i,j] < ymin then ymin:=y[i,j]
                            end;
                        xmax:=x[1];      xmin:=xmax;
                        for j:=1 to npoints do
                            begin
                              if x[j] > xmax then xmax:=x[j];
Scan for max x-value.         if x[j] < xmin then xmin:=x[j]
                            end;
                        yrange:=ymax-ymin;    xrange:=xmax-xmin;
                        writeln;
                        writeln(' ':19,'Graphical Representation of Results');
                        writeln;
                        writeln(' ':20,'xmin = ',xmin:8:3,' ':4,'xmax = ',xmax:8:3);
                        writeln(' ':20,'ymin = ',ymin:8:3,' ':4,'ymax = ',ymax:8:3);
                        writeln;
                        (* set up a line of graph at a time *)
                         ysepar:=0;
                         for ypoints:=1 to maxyscale do
                          begin
                           yaxis:=(maxyscale+1)-ypoints;    ysepar:=ysepar+1;
                           blank:=' ';                      bar:='|';
                           if ysepar=1 then
                             begin
                               blank:='-';  bar:='+';
Set-up the grid                yscale:=ymax-(ypoints-1)*yrange/(maxyscale-1)
                             end;
                           for xpoints:=1 to maxxscale do
                             line[xpoints]:=blank;
                           xpoints:=1;
                           while xpoints <= maxxscale do
                            begin
                             line[xpoints]:=bar;
                             xpoints:=xpoints+10
                            end;
                           (* insert data points now in line[xpoints] *)
                           for k:=1 to nplots do
                            for m:=1 to npoints do
                              begin
                                yvalue:=round(((maxyscale-1)*(y[k,m]-ymin))/yrange+0.5);
                                xvalue:=round(((maxxscale-1)*(x[m]-xmin))/xrange+0.5);
Convert the numbers to positions  if yvalue = yaxis then line[xvalue]:=symbol[k];
in the graph.                 end;
                           if ysepar=1 then
                                  begin
                                    write(yscale:8:3);
                                    for xpoints:=1 to maxxscale do
                                       write(line[xpoints]);writeln
                                  end
                                else
                                  begin
                                    write(' ':8);
                                    for xpoints:=1 to maxxscale do
                                       write(line[xpoints]);writeln
                                  end;
                           if ysepar = 5 then ysepar:=0
                          end;
                        (* x-axis scale calculation *)
                         for i:=1 to 7 do
                           xscale[i]:=xmin+(i-1)*xrange/6;
                         writeln(' ':6,xscale[1]:5:3,' ':15,xscale[3]:5:3,' ':15,
                                 xscale[5]:5:3,' ':15,xscale[7]:5:3);
                         writeln(' ':16,xscale[2]:5:3,' ':15,xscale[4]:5:3,' ':15,
                                 xscale[6]:5:3);
                        end.
```

```
Execution begins...
                  Graphical Representation of Results
                     xmin =    0.100    xmax =     4.000
                     ymin =    0.916    ymax =   109.196

   109.196+---------+---------+---------+---------+---------+---------2
          !         !         !         !         !         !         !
          !         !         !         !         !         !         !
          !         !         !         !         !         !       2 !
    95.661+---------+---------+---------+---------+---------+---------+
          !         !         !         !         !         !         !
          !         !         !         !         !         !         !
          !         !         !         !         !         !  2      !
    82.126+---------+---------+---------+---------+---------+---------+
          !         !         !         !         !         !2        !
          !         !         !         !         !         !         !
          !         !         !         !         !       2 !         !
    68.591+---------+---------+---------+---------+---------+---------+
          !         !         !         !         !      2  !         !
          !         !         !         !         !         !         !
          !         !         !         !         !   2     !         !
    55.056+---------+---------+---------+---------+---------+---------+
          !         !         !         !         ! 2       !         !
          !         !         !         !         !2        !         !
          !         !         !         !       2 !         !         !
    41.521+---------+---------+---------+---------+---------+---------+
          !1        !         !         !       2 !         !         !
          !  1      !         !         !      2  !         !         !
          !     1   !         !         !         !         !         !
          !         !         !         !      !2 !         !         !
    27.98633-33-33-33-33-33-3-33-33-33-33-33-33-2-33-33-33-33-33-33-3-3
          !         !1        !         !     2   !         !         !
          !         !  1!     !         !   2     !         !         !
          !         !   1 1   !         !   2 2   !         !         !
          !         !    1    !         !  2      !         !         !
    14.451+---------+-----11--+---------+22-------+---------+---------+
          !         !       1 1     22  !         !         !         !
          !         !        !1 22      !         !         !         !
          !         !      2 2 22    11 1!        !         !         !
          !      2 22 22  2  !         !  1 11 11 1         !         !
     0.91622-22-2---+---------+---------+--------11-11-11-11-11-11-1-1
         0.100                1.400               2.700               4.000
                  0.750               2.050              3.350

Execution terminated.
```

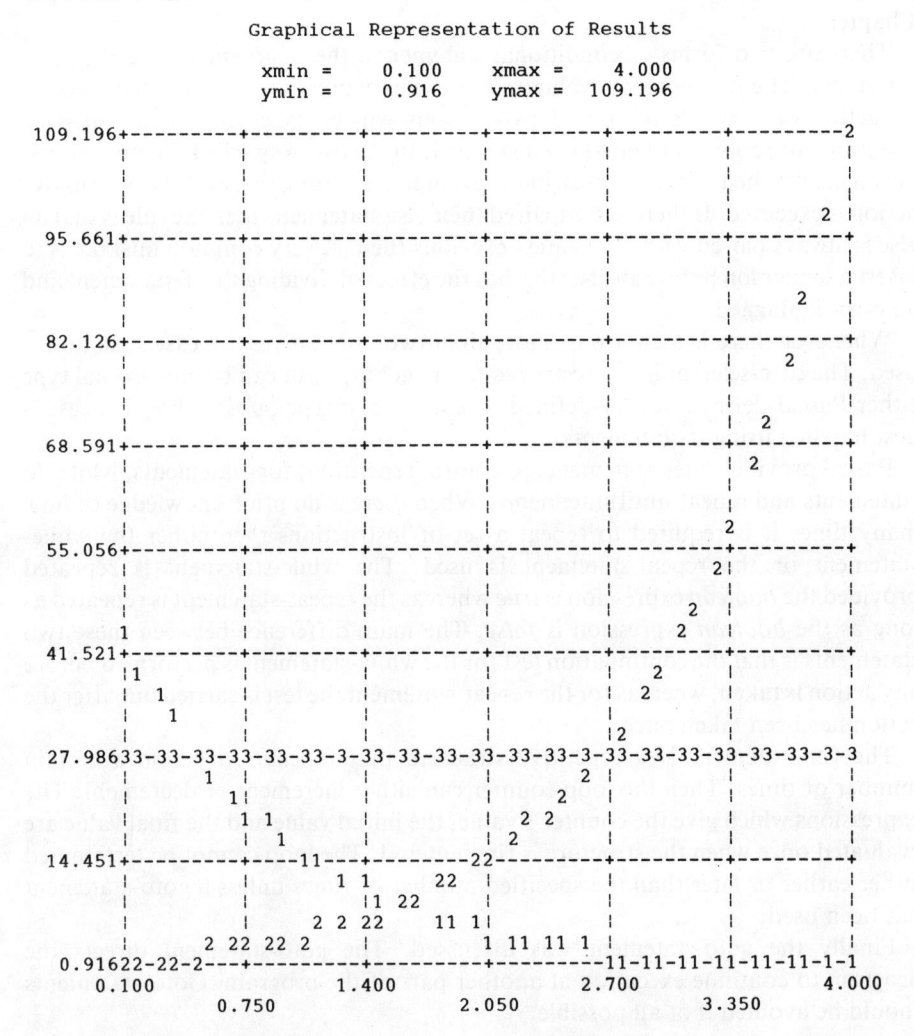

Fig. 4.5 Output from program graph.

Summary

In this chapter the various Pascal statements were examined and particular attention was paid to the structured statements. Structured statements were divided into three types: compound, conditional and repetitive. Structured statements that contain another structured statement of the same type were referred to as nested.

Compound statements are mainly used to group together a number of instructions that we want to consider as one. These are bracketed with **begin** and **end**. Both conditional and repetitive statements involve the boolean expressions discussed in Chapter 3.

There are two forms of conditional statements; the if-statement and the case-statement. The if-statement evaluates a boolean expression to determine whether an action or a choice of one of two actions will be executed. In the one-way selection, an action is either taken or skipped. In the two-way selection, an action is performed if the boolean expression is evaluated as *true* otherwise the alternative action is executed. If there are unpaired **then-else** statements then the rule is that an **else** is always paired with the nearest previous **then**. A very common mistake is to insert a semicolon before an **else**; this has the effect of dividing the if-statement and an error is flagged.

Whenever there is a choice of more than two alternatives the **case** structure is used. The case-selector is not restricted to integer type but can be any ordinal type either Pascal-defined or user-defined. It could be of type *boolean* but this case is best handled using if-statements.

Pascal provides three statements to control repetition; for-statements, while-do statements and repeat-until statements. When there is no prior knowledge of how many times it is required to repeat a set of instructions then either the while-statement or the repeat-statement is used. The while-statement is repeated provided the *boolean* expression is *true* whereas the repeat-statement is repeated as long as the *boolean* expression is *false*. The main difference between these two statements is that the continuation test for the while-statement is performed before any action is taken, whereas for the repeat-statement the test is carried out after the action has been taken once.

The for-statement is a repetitive structure that repeats an action a certain number of times. Then the loop counter can either increment or decrement. The expressions which give the counter's value, the initial value and the final value are evaluated once when the structure is first entered. The loop cannot be terminated either earlier or later than the specified number of times unless a goto-statement has been used.

Finally the goto-statement was discussed. The goto-statement directs the program to continue execution at another part of the program. Goto-statements should be avoided if at all possible.

Problems

4.1 Write a Pascal program to truncate and round a given real number to a prespecified number of decimal places.

4.2 Write a Pascal program to calculate the mean and standard deviation of a given series of real numbers.

4.3 Write Pascal programs to evaluate the following expressions to 7 decimal places. Your program should also indicate the number of iterations needed to achieve such an accuracy.

(i) $\tan^{-1} x = x - \dfrac{x^3}{3} + \dfrac{x^5}{5} - \dfrac{x^7}{7} + \ldots$

(ii) $\ln(1+x) = x - \dfrac{x^2}{2} + \dfrac{x^3}{3} - \dfrac{x^4}{4} + \ldots$

4.4 Write a Pascal program to evaluate $\frac{\pi}{2}$ and $\frac{\pi}{4}$ to 9 decimal places given the series:

(i) $\frac{\pi}{2} = \frac{2}{1} \times \frac{2}{3} \times \frac{4}{3} \times \frac{4}{5} \times \frac{6}{5} \times \frac{6}{7} \ldots$

(ii) $\frac{\pi}{4} = 1 - \frac{1}{3} + \frac{1}{5} - \frac{1}{7} + \ldots$

4.5 Write a Pascal program to tabulate the logarithm to the base 10 of the numbers 3.00, 3.01, 3.02, ... 4.00.

4.6 Write a Pascal program to produce the truth table for

a and (b or not c)

4.7 Write a Pascal program that computes the value of a resistor given the three colour stripes printed on any resistor.

4.8 For the circuit shown in Fig. 4.6 the voltage across the capacitor when the switch is closed at $t = 0$ is given by

$$v_c(t) = V(1 - e^{-\frac{t}{RC}})$$

Write a Pascal program to evaluate $v_c(t)$ for $t = 0$ to $t = 10$ s, given $v = 10$ v, $R = 1\,\Omega$ and $C = 1$ F and hence calculate the rise and delay time for this voltage.

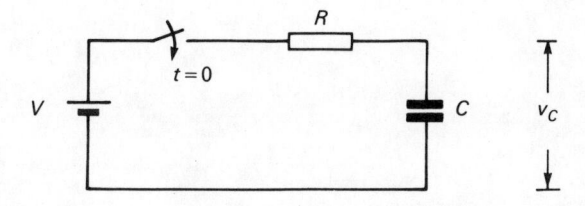

Fig. 4.6 Simple RC network.

4.9 The magnitude and phase of the output voltage for the circuit shown in Fig. 4.7 are given by

$$\left|\frac{V_0}{V_1}\right| = \frac{1}{[(1 - \omega^2 LC)^2 + \omega^2 C^2 R^2]^{\frac{1}{2}}}$$

$$\phi = -\tan^{-1}\left(\frac{\omega CR}{1 - \omega^2 LC}\right)$$

Tabulate and plot the magnitude and phase characteristics for $\omega = 0$ to $\omega = 3$ rad/s.

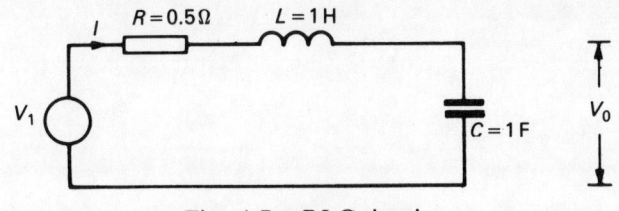

Fig. 4.7 RLC circuit.

4.10 The magnitude and phase of the current I in Fig. 4.7 can be obtained as

$$|I| = \frac{|V_1|}{[R^2 + (\omega L - \frac{1}{\omega C})^2]^{\frac{1}{2}}}$$

$$\phi = \tan^{-1}\left(\frac{\omega L - \frac{1}{\omega C}}{R}\right)$$

Tabulate the magnitude and phase characteristics for $\omega = 0.1$ to $\omega = 10$ rad/s using logarithmic steps.

Functions and Procedures 5

☐ To introduce the concept of writing programs in small modules. **Objectives**
☐ To explain how programmers can write their own functions to supplement those which are available in Pascal.
☐ To introduce procedures.
☐ To discuss the various forms of parameters that can be used with functions and procedures.
☐ To stress the importance of the scope of identifiers.

Why Use Functions and Procedures?

Functions and procedures are basically self-contained subprograms which divide a large Pascal program into smaller sections thus making program writing, reading, debugging and understanding more straightforward. These subprograms cannot be executed on their own but only as a part of a complete Pascal program. Both functions and procedures are declared using the reserved words **function** and **procedure** just after the **var** declarations and before the statement part of a complete Pascal program.

L C T V S

Both functions and procedures are given names (identifiers). The appearance of a subprogram's name in a program is called a subprogram 'call' which instructs the program to execute the subprogram. The main difference between a procedure and a function is that a procedure performs a set of actions, whereas a function returns a single value at the place of its call.

A number of functions provided by standard Pascal have already been discussed. For example, *abs*(x) returns a single real or integer value, depending on the data type of x. Similarly, a number of standard Pascal procedures such as *read*(A,B) which reads two values from the input stream and assign these values to A and B have been discussed. Note that a procedure call *read*(A,B) is a statement in its own right whereas a function call must form a part of an expression or statement e.g. y: = *abs*(x); a: = *cos*(y) + *sin*(y);

There are a number of very good reasons why the use of these two types of subprogram should be learned. The reasons become more apparent as larger and larger programs are written:

(i) Each subprogram can be written and tested separately, which means that each subprogram can be read and understood separately.
(ii) A number of subprograms can be used to build a large program thus making the writing of large programs easier.
(iii) Once written, a subprogram can be called a number of times within a program.
(iv) Once written and thoroughly tested, a subprogram can be used in another Pascal program. Most programmers tend to write subprograms which they put together to form another complete program.

(v) Subprograms are given names which aid immensely in the understanding of a program.

(vi) Subprograms are self-contained, therefore replacing them with an improved or updated version is straightforward.

A good programmer 'divides and rules', *i.e.* subdivide the given problem into smaller steps. This method of writing programs is called 'modular development' or 'top-down' approach. The top-down approach initially describes the problem at the most general level and then tries to break down the problem into smaller sub-problems. The fundamental advantage of programming this way is that it enables a programming problem to be solved in small steps without losing sight of the main objective. The breakdown of a large program is achieved with the aid of subprograms, and since subprograms may contain declarations of other subprograms this enables the subdivision into suitably small sections. The result is that a program, however large, becomes relatively easy to write. A program which is long and monolithic usually frustrates the reader. If a subprogram makes the program more readable it has achieved one of its aims. Even if the subprogram is only called once it has earned its place in a program.

The importance of functions and procedures cannot be overemphasised.

See Cooper, D. and Clancy, M. *Oh! Pascal* (W.W. Norton, 1982).

Functions

Since functions are easier to understand than procedures they are discussed first. As mentioned earlier, a function is a subprogram which returns a single value when called. A number of standard Pascal functions have been introduced in chapters 2 and 3. All these functions have a single argument of *real* or ordinal type and yield a result of an appropriate data type (See Appendix B). The difficulty of not having standard Pascal functions to evaluate the logarithm to the base 10, exponentiation, and so on, has been encountered. The method for creating functions is now discussed.

Since a function call returns a single value, this must be declared to be of some specific data type. Also the function's name must be assigned to an expression which calculates the function's value. Consider a function which evaluates the hyperbolic sine of a real variable x:

function sinh(x:*real*):*real*;   ⬅——— function-heading
(∗ calculates hyperbolic sine ∗)
begin
  sinh: = 0.5 ∗ (*exp*(x) − *exp*(−x))   ⬅——— function-body
end (∗ sinh ∗);

A good practice is to write the name of a subprogram as a comment after the subprogram's last **end**.

As can be seen from the above 'function-declaration' the structure of a function is very similar to a complete Pascal program. The reserved word **function** is used instead of **program** in the 'function-heading' and a semicolon is used rather than a period to terminate the function declaration. Following the reserved word **function**, the name (function-identifier) of the function is given, in this case sinh. The main purpose of the function-heading is to name the function (sinh) and to declare the type of result that is obtained from the evaluation of the function (*real*).

After the function-identifier, and enclosed in parentheses, there is a variable declaration list which is referred to as the 'formal-parameter' list or 'dummy parameters'. This list describes the variables which form the input data to the function and also their type. In the given example there is only one variable x which is declared to be of type *real*. The formal-parameter list is written exactly as the list of declarations for the **var** section of a Pascal program but without the reserved word **var**. Examples of function-heading are:

(i)   **function** accept(n:*real*;min,max:*integer*):*boolean*;
(ii) **function** log(a:*real*;n:*integer*):*real*;
(iii) **function** enigma(c:*char*):*char*;
(iv) **function** even(n:*integer*):*boolean*;

In general, there can be any number of identifiers in the formal-parameter list of simple type, of a previously structured type, or of pointer type (see Chapter 6). The type of value resulting from the evaluation of the function can be of scalar, subrange or pointer type but **not** array type since the function returns only a single value.

It is possible to have functions without a formal-parameter list but these have limited use in well-written programs.

Following the function-heading there is a sequence of statements which form the 'function-body' which describe the operations to be performed to obtain the desired result. The function-body must be bracketed with **begin-end**. Also in the function body there must be at least one statement which assigns a value, of result-type, to the function's name identifier. Execution of this statement defines the value of the function. In the sinh example there is only one statement.

With the knowledge so far gained a general form for a function can be written as follows:

```
( * function heading * )
function   function-identifier(formal-parameter list):result-type;
( * function body * )
begin
   statements
   statements
   function-identifier: = ......
end ( * function-identifier * );
```

Shown side-by-side are a function declaration that calculates the hyperbolic cosine and a complete program which performs the same task:

```
function cosh(x:real):real;          program HyperbolicCos(input,output);
begin                                var cosh,x:real;
   cosh: = 0.5 * ((exp(x) + exp( - x))   begin
end ( * cosh * );                       read(x);
                                        cosh: = 0.5 * ((exp(x) + exp( - x));
                                        write(cosh)
                                     end.
```

It is important to note the function cosh cannot be executed on its own and that no actual computation is performed without involving a function call. The complete Pascal program incorporating the function cosh can be written as:

```
program HyperbolicCos(input,output);
var a:real;
function cosh(x:real):real;
( * hyperbolic cosine function * )
```

The type of result of a function can never be a structured type. Note no **var** in the formal parameter list.

```
        begin
            cosh: = 0.5 * ((exp(x) + exp(-x))
        end (* cosh *);
    (* program statement part *)
    begin
        readln(a); writeln(cosh(a))
    end.
```

In this program the function cosh is invoked (called) by writing the function-identifier. The effect of a function call is to execute the statements in the function-body, in this case the evaluation of an expression and the assignment of this expression to the function-identifier. The function call must supply an 'actual' parameter or 'calling' parameter for each formal parameter in the formal-parameter-list. In the above example the actual parameter is a. In general, a function has a number of variables in the formal parameter-list. For example consider the function heading:

Note no **var**.

    function Sample(a,b:*real*;i:*integer*;c:*char*):*boolean*;

A function call can be:

    if Sample(3.5E−2,BB,5,'C') **then** ......
    (* assume that BB has been declared as real *)

The actual parameters (and in this case some actual numbers) are paired-off from left to right with the formal parameters according to their position and type *i.e.* 3.5E−2 ⟶ a, BB ⟶ b, 5 ⟶ i, and 'C' ⟶ c. The formal parameters are used only as place and type indicators *i.e.* when the function is called in effect there is:

    a: = 3.5E−2;   b: = BB;   i: = 5;   c: = 'C';

The actual parameters appear on the right hand side of the assignment statement and the formal parameters on the left hand side.

Worked Example 5.2   Write Pascal functions to:
    (i)   raise an *integer* variable to a *real* power
    (ii) calculate the tangent of an argument given in radians
    (iii) test for even numbers
    (iv) calculate the logarithm to the base 10 of a given number.

Solution:

(i) **function** power(n:*integer*;a:*real*):*real*;   (ii) **function** tan(x:*real*):*real*;
    begin   **begin**
       power: = *exp*(a * *ln*(n))   tan: = *sin*(x)/*cos*(x)
    end (* power *);   **end** (* tan *);
(iii) **function** even(n:*integer*):*boolean*;   (iv) **function** log(x:*real*):*real*;
    begin   **begin**
       even: = (**not** *odd*(n))   log: = *ln*(x)/*ln*(10)
    end (* even *);   **end** (* log *);

Write a Pascal program to calculate the transmission parameters of a telephone line given R = 55 Ω/km, L = 48 mH/km, G = 0.6 μS/km and C = 0.04 μF/km.

Worked Example 5.1

Solution:

The equivalent circuit of an incremental length $\triangle x$ of a transmission line is shown in Fig. 5.1 where R is the resistance of the line per unit length, L is the inductance of

See Lathi, B.P. *Signal System and Communication* (Wiley, 1965).

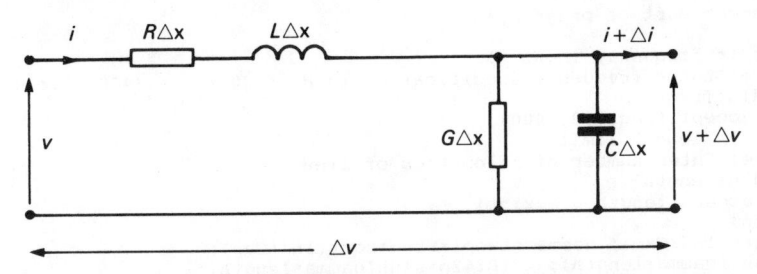

Fig. 5.1 Equivalent circuit of a transmission line.

the line per unit length, G is the conductance of the line per unit length and C is the capacitance of the line per unit length.

The characteristic impedance Z_0 and propagation constant γ for the line in the speech band can be calculated as:

$$Z_0 = \sqrt{\frac{L}{C}} \quad \text{and} \quad \gamma = j\omega\sqrt{LC}$$

Since concern is with the calculation of results within the speech band then it is necessary to ensure that the input frequency entered in the program lies betwen 300 Hz and 3.4 kHz. A useful function can be written to test if the input frequency lies within this limit as:

> **function** accept (f:*real*;min,max:*integer*): *boolean*;
> (* accepts a value within a min and max limit *)
> **begin**
>   accept: = (f > = min) **and** (f < = max)
> **end** (* accept *);

It may be thought that this function is an overkill for such a small program, but you should find this type of function very useful when writing larger programs.

It is seen how this *boolean* function could help clarify the flow of a program.

```
program TwoPortTransmParameters(input,output);

(* This program calculates the twoport transmission
   parameters of a telephone line in the speech band *)

const  Ind=48E-3;   Cap=0.04E-6;
       twopi=6.28318539778;
var    Zo,gamma,freq,length,A,B,C,D:real;

function accept (f:real;min,max:integer):boolean;
(* accepts values between a min and a max *)
begin
   accept:=( (f>=min)   and   (f<=max)  )
end  (* accept *);
```

A good habit to adopt is to leave a blank line between subprogram definitions as shown.

```
          function cosh(x:real):real;
          (* hyperbolic cosine *)
          begin
            cosh := 0.5*(exp(x)+exp(-x))
          end (* cosh *);

          function sinh(x:real):real;
          (* hyperbolic sine *)
          begin
            sinh:=0.5*(exp(x)-exp(-x))
          end (* sinh *);

(* statement part of program *)
begin
  repeat (* frequency input *)
    write('Enter frequency specification in Hz (300 < f < 3400) ');
    readln(freq)
  until accept(freq,300,3400);
  repeat (* length limit *)
    write('Enter number of kilometres of line =');
    readln(length)
  until accept(length,0,maxint);
  writeln;
  Zo:=sqrt(Ind/Cap); gamma:=twopi*freq*sqrt(Ind*Cap);
  A:=cosh (gamma*length);    B:=Zo*sinh(gamma*length);
  C:=sinh(gamma*length)/Zo;  D:=A;
  writeln('| A     B |',' ':5,'|',A:7:3,' ':2,B:7:3,'|');
  writeln('|         |',' ':3,'= |',' ':20,'|');
  writeln('| C     D |',' ':5,'|',C:7:3,' ':2,D:7:3,' '|')
end.

Execution begins...

Enter frequency specification in Hz (300 < f < 3400) 1000
Enter number of kilometres of line =20

| A     B |    |123.122   134869.131|
|         | =  |                    |
| C     D |    |  0.112      123.122|

Execution terminated.
```

Local Declarations within Functions

In the previous examples the result of a function was obtained by obeying a single statement and using only variables that were given in the formal-parameter list. But in the majority of cases a number of statements are required to calculate the result of a function using more identifiers than the ones provided in the formal-parameter list. These extra identifiers are useful only to the function itself *i.e.* there are 'local' identifiers. Such identifiers could be declared in the declaration section of the complete program and in this case they are called 'global' identifiers. However by doing so we make the function dependent on data outside itself, *i.e.* it is no longer independent of the main program. To overcome this problem Pascal allows functions (and procedures) to have local declarations which are made between the function-heading and the function-body. These can be **L**abels, **C**onstants, user defined **T**ypes, **V**ariables and/or other **S**ubprograms. These local definitions and declarations are one of the most important and unfortunately most-abused feature of Pascal.

Before proceeding any further consider the following points:
(i) Subprograms are easier to write and maintain if they are self-contained.
(ii) Identifiers that are defined or declared in the main program are called

'global' identifiers and can be used anywhere in the program.

(iii) Identifiers that are defined within a subprogram are called 'local' identifiers and have no definition outside the subprogram.

(iv) The recently-accepted Pascal standard requires that for-statements use local variables in the function (and procedure) in which they occur, so that the loop does not accidentally change the value of a global variable.

(v) Memory space is allocated for local identifiers when the subprogram is called and released when execution of the subprogram is completed. Therefore declaring local identifiers makes more efficient use of memory space.

To help clarify the concept of local and global identifiers consider a program to calculate the factorial of a given integer number. Bearing in mind that it is required to write the program in a modular form then two functions can be used, one to calculate the factorial and another to limit the value of the given integer so that its factorial value does not exceed the integer limit of our computer. For the VAX 11/750 this limit is 12.

Wilson, I.R. and Addyman, A.M. *A Practical Introduction to Pascal-with BS6192* (MacMillan, 1982).

12! = 479001600

```
program FactorialTest(input,output);

(* Prints factorials of numbers   *)

var number:integer;

function factorial(n:integer):integer;
type posinteger= 0 .. maxint;
var i,f:posinteger;
begin
   f:=1;
   if n=0 then factorial:=1
          else
               begin
                  for i:=n downto 1 do
                     f:=f*i;
                  factorial:=f
               end
end (* factorial *);

function accept(n,min,max:integer):boolean;
(* accepts integer values within a min and max limit *)
begin
   accept:=(  (n>=min) and (n<=max)   )
end (* accept *);

begin
   repeat (* 0<=number<12 *)
     write('Enter positive integer =');readln(number)
   until accept(number,0,12);
   writeln('Factorial of ',number:1,' = ',(factorial(number)):1);
   writeln
end.

Execution begins...

Enter positive integer =10
Factorial of 10 = 3628800

Execution terminated.
```

The program could also have been written as given below. Its main drawback is that the **function** factorial is not independent but relies on declarations made globally in the main program. Also, according to the accepted Pascal standard we should have used local variables for the for-statement. The following program is a typical example of how *not* to write Pascal programs:

Apart from global/local variable problems there are a few more, can you spot them?

```pascal
program Factorial2(input,output);

(* Displays factorials of numbers  *)
type posinteger = 0 .. maxint;
var number,i,f:posinteger;
function factorial(number:posinteger):posinteger;
begin
 f:=1;
 if number=0 then factorial:=1
       else
         begin
           for i:=number downto 1 do
             f:=f*i;
             factorial:=f;
         end
end (* factorial *);

begin
 repeat (* 0<=n<12*)
  write('Enter positive integer =');readln(number);
 until (  (number>=0) and (number<=12)  );
 writeln('Factorial of ',number:1,' = ',(factorial(number)):1);
end.

Execution begins...

Enter positive integer =10
Factorial of 10 = 3628800

Execution terminated.
```

Therefore, the general structure of a function, which is very similar to a Pascal program structure, can be written as:

(* function heading *)
function function-identifier (formal-parameter list):result-type;
(* local function declarations *)
label
const
type
var
procedure and **functions**
(* function body *)
begin
   ....
   ....
function-identifier: = ...
end;

Worked Example 5.3   Write a Pascal program to calculate the tangent of an angle given in degrees.

Solution:

This program can be written by incorporating another function to convert degrees to radians in the function tan given in Example 5.1.

```pascal
program Tangent(input,output);

(* calculates the tangent of an angle *)
```

```
        var angle:real;
            yes:char;

        function YesNo(YN:char):boolean;                    Another useful function.
         begin
           YesNo:=( (YN='Y') or (YN='y') )
         end (* YesNo *);

        function tan(degrees:real):real;
         var theta:real;
          function DegreesToRadians(theta:real):real;
            const pi=3.141592653589;
            begin
              DegreesToRadians:=(pi/180)*theta
            end (* tan *);

          begin
            theta:=DegreesToRadians(degrees);
            tan:=sin(theta)/cos(theta)
          end (* tan *);

        begin
          repeat (* another ? *)
            write('Enter angle in degrees = ');readln(angle);
            writeln('tan(',angle:1:2,') = ',tan(angle):1:2);
            write('Type y for another angle = ');readln(yes);writeln;
          until not YesNo(yes)
        end.

Execution begins...

Enter angle in degrees = 60
tan(60.00) = 1.73
Type y for another angle = y

Enter angle in degrees = 45
tan(45.00) = 1.00
Type y for another angle = n

Execution terminated.
```

Write a program that calculates the input resistance of a resistive ladder network. **Worked Example 5.4**
Use procedures for the calculation of the parallel and series evaluations.

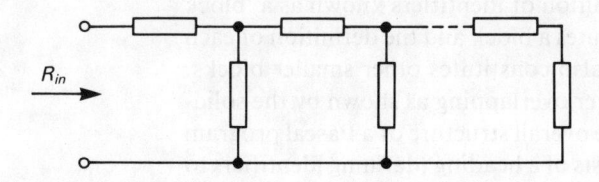

```
program resistiveladder3(input,output);

(* evaluates a resistive ladder using functions *)

const maxmeshsize = 10;

type resarray = array [1..maxmeshsize] of real;

var mesh, meshsize : integer;
    rtemp : real;
    rs, rp : resarray;

function parallel(x, y:real):real;
begin
```

85

```
    parallel := x * y / (x + y)
end; (* function parallel *)
function series(x, y:real):real;
begin
  series := x + y
end; (* function series *)

begin (* main program *)
  repeat
    write('Enter number of meshes (1 - ',maxmeshsize:3,' ): ');
    readln(meshsize);
  until ((meshsize > 0) and (meshsize < maxmeshsize));
  for mesh := 1 to meshsize do
  begin
     write('Enter parallel and series resistors for mesh ',
           mesh:3,': ');
     readln(rp[mesh], rs[mesh])
  end;
  rtemp := series(rp[1],rs[1]);
  mesh := 1;
  while (mesh < meshsize) do
  begin
    mesh := mesh + 1;
    rtemp := parallel(rtemp, rp[mesh]);
    rtemp := series(rtemp,rs[mesh])
  end;
  writeln('Equivalent resistance is ',rtemp:12:4)
end. (* program resistiveladder3 *)

Execution begins...

Enter number of meshes (1 -  10 ): 3
Enter parallel and series resistors for mesh    1: 1 1
Enter parallel and series resistors for mesh    2: 1 1
Enter parallel and series resistors for mesh    3: 1 1
Equivalent resistance is       1.6250

Execution terminated.
```

Scope of Identifiers and Side Effects

Pascal uses a convention regarding the definition of identifiers known as a 'block structure'. The main Pascal program constitutes a block and the definition of each function and procedure within a program also constitutes other smaller blocks. Some of these blocks may be nested but never overlapping as shown by the solid-line boxes of Fig. 5.2. Therefore consider the overall structure of a Pascal program as a set of blocks. Syntactically a block consists of a heading (defining identifiers to be used for communication), definitions and declarations (**L, C, T, V, S**), if any, and the statement part. When thinking in terms of blocks, there is no real distinction between a program and subprograms.

As can be seen from the skeleton program shown in Fig. 5.2 any of these blocks may introduce new identifiers in its declaration part. Identifiers declared inside a subprogram (local identifiers) have no meaning outside that particular subprogram. The *scope* of an identifier can be referred to as that section of program in which it can be used *i.e.* the block where it is defined. The rules of identifier scope in Pascal are:

(1) The scope of an identifier is the block in which the declaration or definition is done, including all the blocks enclosed by that block.

> Make sure the scope of identifiers is understood.

(2) No identifier may be declared more than once within the same block. If this is done, we have an ambiguity.

(3) An identifier may be used only within the block where it is declared.

(4) If the same identifier is declared both in an inner block (local) and in an outer block then the most local identifier takes precedence *i.e.* assignment, within a subprogram, to a local variable does not change the value of the global variable with the same name.

The scope of various identifiers given in the skeleton program of Fig. 5.2 are shown with solid-lines. The scope of global identifiers is the entire program, whereas the scope of local identifiers is limited to the block where they are declared. Note that function-identifiers and procedure-identifiers are also subject to the same rules of scope as other identifiers.

Consider a number of points which can be made with reference to Fig. 5.2:

(i) The function-identifier Valid can only be used within the block structure defined by the function Test. Outside this block Valid has no meaning.

(ii) The identifier theta is declared both as a global variable and also as a local variable within the function Phase. This conflict is solved using rule (4) given above. The global variable theta becomes inaccessible and retains whatever value it had when the local variable theta was declared. Even though the program would operate correctly, this is very poor programming style and should be avoided.

(iii) The variables x, y and freq are declared in both function Modulus and Phase. In this case there is no conflict since these identifiers are valid only within the block in which they are declared. As far as the compiler is concerned they are different identifiers.

(iv) All the subprograms defined are independent of the main program except function Phase, since the result of this function relies on a type definition performed globally *e.g.* posinteger.

From the above it can be seen that extreme care is required when writing subprograms. There is one more sin which may be committed if care is not exercised: changing the value of a global variable during the evaluation of a function. This is called a 'side-effect'. Creation of functions which have side-effects should be avoided; the accidental change of global values in a subprogram causes a lot of headaches to say the least. An example of a side effect is shown in Fig. 5.2 in the function Modulus: 'f1: = 1E−20;'.

> For large programs global parameter passing is an acceptable practice with procedures.

Before leaving this section let us state some important points:

(1) A program is strictly modular if its subprograms are independent of each other and of the main program.

(2) All data values declared at the beginning of a block can be used anywhere within that block including any inner blocks but in no other.

(3) Use of local identifiers minimises the memory storage requirement since they exist only while they are called.

(4) Local variables provide a means of protecting data from undesirable accessing.

(5) Because constants cannot be corrupted, the positioning of **const** declarations is not critical.

(6) There is no provision for subprogram libraries in standard Pascal.

> A major drawback of Pascal is the lack of provision for libraries.

```
program ActiveFilter(input,output);
const pi = 3.14159265359;
type posinteger = 1 .. maxint;
     xarray = array[1..20] of real;
var  i,j,n,k : posinteger;
     theta,f1,frequency:real;
     magnitude:xarray
     ..................

    procedure Readata;
    begin
       ............
    end (* Readata *);

    procedure NewLine(x:integer);
    var j:integer;
    begin
       ............
    end (* NewLine *);

    function Modulus(x,y,freq:real):real;
    begin
       ............
       f1: = 10E – 20;
       Modulus: = ........
    end (* Modulus *);

    function Phase(x,y,freq:real):real;
    var theta,angle:real;
        k,l: posinteger;
    begin
       ............
       angle: = arctan(theta);
       Phase: = ......
       ............
    end (* Phase *);

    function Test(a,b:integer;w:real):boolean;
    var qw:char;

        function Valid(c:char):boolean;
        begin
           ............
        end (* Valid *);

    begin
       ............
    end (* Test *);

(* main program body *)
begin
   ............
   ............
end.
```

Fig. 5.2   Skeleton program to illustrate block structure.

Procedures

Functions are subprograms which produce a single result immediately on return from a function call. However, in the majority of cases subprograms are required to return several results (some of **array** type) or no results at all but require a certain action to be performed. Such subprograms can be written in Pascal using **procedure**. Unlike a function call, a procedure call does not itself return a single value in place of its procedure-identifier; instead it executes a number of statements which may or may not produce results. In general, procedures are used more frequently than functions to create subprograms.

Procedure declarations have similar structure to a program and therefore to a function:

```
                                                            procedure
procedure procedure-identifier (formal parameter list);  ← heading
declarations label
             const
             type
             var                                          ← local        L C T V S
             procedure and function                          declarations
begin
     .............
     .............                                        ← procedure
     .............                                          body
end (* procedure-identifier *);
```

Procedure declarations, like function declarations are inserted after the **var** declarations of the main program. The main differences between a function and a procedure declaration are:

(i) The reserved word **procedure** is used instead of **function**.
(ii) The parameters in the formal parameter list of procedures can be used:
  a) to supply values for use in the procedure (value-parameters), as in the case of functions;
  b) to return results from the procedure (variable-parameters or reference-parameters);
  c) a combination of *a*) and *b*).
(iii) Result-type for the procedure itself is not specified since there is no explicit return of result.
(iv) There is no assignment statement (anywhere) to assign a value to a procedure-identifier.

Procedures can declare local variables with their scope as discussed in the previous section. There are also procedures, as in the case of functions, with no formal parameter-list or with functions/procedures as formal parameters. Procedures therefore can have the following as formal parameters:

(i)   no formal parameters
(ii) value parameters
(iii) variable parameters
(iv) procedures and/or functions
(v)   a combination of the above.

An error is flagged if a procedure-identifier is used as part of a statement; procedure-identifiers are statements in their own right.

Remember that a procedure call is the same as a function call, except that a procedure call is a statement in itself. The association between formal and actual parameters is exactly the same as for functions.

Procedures with No Formal Parameters

If a procedure is required to perform a predefined action *e.g.* to draw a line across the output or to temporary suspend the output, we can achieve this by using procedures with no formal parameters. For example:

```
procedure DrawLine;
const width = 80;
var i:integer;
begin
  for i:=1 to width do
    write('-');writeln
end (* DrawLine *);
```

```
procedure Pause;
begin
  write('Press return to continue');
  readln
end (* Pause *);
```

As can be seen from the above two examples, useful as they are, these procedures have very limited capabilities, especially the DrawLine procedure. This procedure always draws an 80 character line using hyphens. A well written procedure allows greater flexibility by using formal parameters.

Procedures with Value Parameters

As discussed previously, the formal parameter list defines a set of dummy variables (formal parameters) which are replaced by the corresponding actual parameters (value parameters) when a function is called. Similarly value-parameters can be used to pass values to a procedure. For example the DrawLine procedure can be re-written as:

```
procedure DrawLine (width:integer;symbol:char);
var i:integer;
begin
  for i:=1 to width do
    write(symbol);writeln
end (* DrawLine *);
```

A procedure call 'DrawLine(30,'*');' prints 30 stars and a procedure call 'DrawLine(15,'-');' draws a line with 15 hyphens. The DrawLine procedure is useful if it is required to print graphs or underline output. The following two procedures, which are very simple, can also be used to help with program output:

```
procedure Space(number:integer);
var i:integer;
begin
  for i:=1 to number do
    write(' ')
end (* Space *);
```

```
procedure NewLine(number:integer);
var i:integer;
begin
  for i:=1 to number do
    writeln
end (* NewLine *);
```

As an exercise using the above procedures try to *figure-out* the output of the following program:

> **program** Sigma(*output*);
> **procedure** Space(number:*integer*);
>   ......(* as above *) ......
> **procedure** NewLine(number:*integer*);
>   ......(* as above *) ......
> **procedure DrawLine** (number:*integer*;symbol:*char*);
>   ......(* as above *) ......
> **begin**
> Space(20);DrawLine(4,' – ');NewLine(1);
> Space(21);DrawLine(1,' \ ');NewLine(1);
> Space(22);DrawLine(1,' \ ');NewLine(1);
> Space(21);DrawLine(1,'/');NewLine(1);
> Space(20);DrawLine(1,'/');NewLine(1);
> Space(20);DrawLine(4,' – ');NewLine(4)
> **end.**

Write a general purpose plotting procedure. **Worked Example 5.5**

Solution:

This procedure is based on the program given in Case Study 2 of Chapter 4. Procedure plot has four formal parameters:

x = a one dimensional array, which forms the abscissa of the graph
y = a two dimensional array, which contains the ordinate values for the various curves
Points = an integer, which is the number of points to be plotted.
Plots = an integer, which is the number of curves to be plotted on the graph.

```
procedure plot(x:onedarray; y:twodarray; points,plots:posinteger);
const maxxscale = 61;
      maxyscale = 21;
var ymax,ymin,xmin,xmax,yrange,xrange,yscale :real;
    ysepar,ypoints,yaxis,xpoints,yvalue,xvalue :posinteger;
    i,j,k,m :posinteger;
    blank,bar :char;
    symbol :array [1..9] of char;
    line   :array [1..100] of char;
    xscale :array [1..12] of real;
begin
    for i := 1 to plots do
        symbol[i] := chr(i + ord('0'));
    ymax := y[1,1];
    ymin := ymax;
    for i := 1 to plots do
        for j := 1 to points do
        begin
            if y[i,j] > ymax then
                ymax := y[i,j];
            if y[i,j] < ymin then
                ymin := y[i,j];
        end;
    xmax := x[1];
    xmin := xmax;
    for j := 1 to points do
```

```pascal
        begin
            if x[j] > xmax then
                xmax := x[j];
            if x[j] < xmin then
                xmin := x[j];
        end;
    yrange := ymax - ymin;
    xrange := xmax - xmin;
    writeln(' ':19,'Graphical Representation of Results');
    writeln;
    writeln(' ':20,'xmin = ',xmin:8:3,' ':4,'xmax = ',xmax:8:3);
    writeln(' ':20,'ymin = ',ymin:8:3,' ':4,'ymax = ',ymax:8:3);
    writeln;
    ysepar := 0;
    for ypoints := 1 to maxyscale do
    begin
        yaxis := (maxyscale+1)-ypoints;
        ysepar := ysepar + 1;
        blank := ' ';
        bar := '|';
        if ysepar = 1 then
        begin
            blank := '-';
            bar := '+';
            yscale := ymax -(ypoints -1)*yrange/(maxyscale-1)
        end;
        for xpoints := 1 to maxxscale do
            line[xpoints] := blank;
        xpoints := 1;
        while xpoints < maxxscale do
        begin
            line[xpoints] := bar;
            xpoints := xpoints + 10;
        end;
        for k := 1 to plots do
            for m := 1 to points do
            begin
                yvalue := round(((maxyscale-1)*(y[k,m]-ymin))/yrange+0.5);
                xvalue := round(((maxxscale-1)*(x[m]-xmin))/xrange+0.5);
                if yvalue = yaxis then
                    line[xvalue] := symbol[k]
            end;
        if ysepar = 1 then
        begin
            write(yscale:8:3);
            for xpoints := 1 to maxxscale do
                write(line[xpoints]);writeln;
        end else begin
            writeln(' ':8);
            for xpoints := 1 to maxxscale do
                write(line[xpoints]);writeln;
        end;
        if ysepar = 5 then
            ysepar := 0;
    end;
    for i := 1 to 7 do
        xscale[i] := xmin + (i-1)*xrange/10;
    writeln(' ':6, xscale[1]:5:3,' ':14,xscale[3]:5:3,' ':14,
        xscale[5]:5:3,' ':14,xscale[7]:5:3);
    writeln(' ':16, xscale[2]:5:3,' ':14,xscale[4]:5:3,' ':14,
        xscale[6]:5:3);
end;
```

The output from the above program is identical to the one given in Fig. 4.5.

The calling statement: 'plot(xaxis,ycurves,npoints,nplots);' has four value-parameters with types that correspond to the formal parameter types. Note that **procedure** plot is dependent on the main program for its array declarations. To avoid this some Pascal implementations allow the procedure heading to be written as:

```
procedure plot(x:array[1..100] of real;
               y:array [1..9,1..10] of real;Points,Plots:integer);
```

But it must be stressed this is *not* standard Pascal. According to the standard (BS6192) array type can only be specified as an identifier and not as an explicit definition shown above.

Wilson, I.R. and Addyman, A.M. *A Practical Introduction to Pascal-with BS6192* (MacMillan, 1982).

Using Global Variables

So far how formal parameters are used to supply values to a subprogram from the calling program has been discussed. Suppose now that it is wished to return values to the calling program from the subprogram rather than just performing an action *i.e.* plotting a graph. There are two methods for communicating values back to the calling program: (*1*) by modifying global variables in a procedure and (*2*) by using variable parameters in the procedure formal parameter list.

Provided that, within the procedure, there is no local variable with the same name as the global variable then a procedure can be used to assign values to global variables. These can then be accessed by other subprograms or the main program. The main drawback of this method is that it is prone to side-effects which are difficult to detect. There are three situations in which global parameter passing is often used: (1) to read data into a program (2) to manipulate common data by various subprograms. (3) to preserve some variable values from one call of a subprogram to the next. This cannot be done with local variables.

The following exercise has very simple calculations but it highlights all the above three points.

Calculate the effective inductance of two coils connected (a) in series and (b) in parallel.

Worked Example 5.6

Solution:

The series and parallel connections of the coils are shown in Fig. 5.3. The effective series and parallel inductance can be calculated from:

Fidler, J.K. *Introductory Circuit Theory* (McGraw Hill, 1980).

$$L_{\text{series}} = L_1 + L_2 + 2M$$
$$L_{\text{parallel}} = \frac{L_1 + L_2 + \sqrt{M}}{L_1 + L_2 - 2M}$$

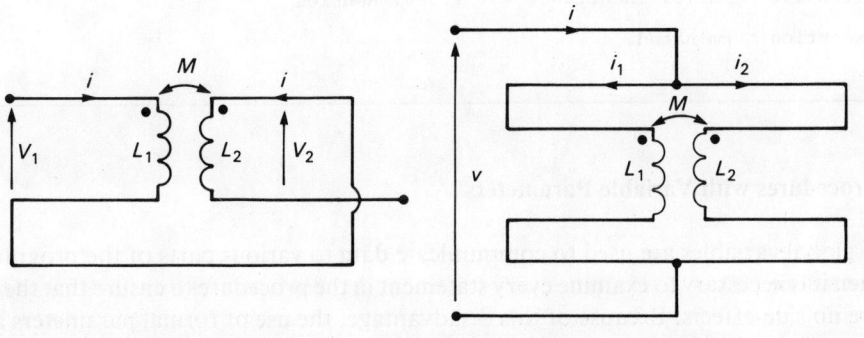

Fig. 5.3 Inductors connected in series and parallel.

A Pascal program using the above equations can be written as:

```pascal
program Inductors(input,output);

(* Calculates interaction between two inductors
   in series and parallel realizations           *)

var L1,L2,L,M,K:real;
    AorO:char;
function accept(test,c1,c2:char):boolean;
begin
  accept:=( (test=c1) or  (test=c2) )
end (* accept *);

procedure readata;
begin
 write('Enter value of first inductor in Henries =');readln(L1);
 write('Enter value of second inductor in Henries =');readln(L2);
 write('Enter coupling coefficient =');readln(K);
 repeat (* 'A' or 'O' *)
    write('A(iding or O(pposing currents =');readln(AorO);
    writeln
 until accept(AorO,'A','O');
 M:=K*sqrt(L1*L2)
end (* readata *);

procedure  series;
begin
 if AorO='A' then L:=L1+L2+2*M
             else L:=L1+L2-2*M
end (* series *);

procedure parallel;
begin
 if AorO='A' then L:=(L1*L2+sqrt(M))/(L1+L2+2*M)
             else L:=(L1+L2+sqrt(M))/(L1+L2-2*M)
end (* parallel *);

begin
 readata;
 writeln;
 series;
 writeln('Effective series inductance = ',L:8:4,' Henries');
 parallel;
 writeln('Effective parallel inductance = ',L:8:4,' Henries')
end.

Execution begins...

Enter value of first inductor in Henries =1.2
Enter value of second inductor in Henries =3.4
Enter coupling coefficient =.2
A(iding or O(pposing currents =O


Effective series inductance =    3.7920 Henries
Effective parallel inductance =    1.3807 Henries

Execution terminated.
```

Procedures with Variable Parameters

If global variables are used to communicate data to various parts of the program then it is necessary to examine every statement in the procedure to ensure that there are no side-effects. Because of this disadvantage, the use of formal parameters as variable parameters is preferred. This can be achieved by preceding the formal parameters by the reserved word **var**. This declaration must contain only **var**iables; it cannot contain **const**ants **procedure**s, **function**s or any other expressions. For example a procedure-heading can be written as:

   procedure InputOutput(A,B:*real*;**var** C,D:*real*);

and a procedure call as

   InputOutput(AA,BB,CC,DD);

The above procedure call specifies that the AA and BB actual parameters are value parameters and that the CC and DD actual parameters are variable parameters. Also the actual parameters (AA,BB,CC,DD) replace the formal-parameters (A,B,C,D) with respect to their position. Consider two examples which have identical formal parameters:

Use of formal parameters can become quite cumbersome if a number of variables are passed between subprograms using this method.

   procedure CompMult (a,b,c,d:*real*;**var** re,im:*real*);
   (∗ complex number multiplication ∗)
   begin
      re: = (a∗c)−(b∗d);
      im: = (c∗b)+(a∗d)
   end (∗ CompMult ∗);

   procedure CompDiv(a,b,c,d:*real*;**var** re,im:*real*);
   (∗ complex number division ∗)
   var denom:*real*;
   begin
      denom: = (c∗c+d∗d);
      re: = ((a∗c)+(b∗d))/denom;
      im: = ((c∗b)−(a∗d))/denom
   end (∗ CompDiv ∗);

Since Pascal does not support complex number arithmetic it is necessary to write procedures to manipulate complex numbers by treating the real and imaginary parts as different entities. For example, the above procedures divide or multiply two complex numbers of the form (a + jb) and (c + jd). The results are returned to the main program using the formal-parameters re(for the real part) and im (for the imaginary part). It is worth restating here that the formal-parameters can be considered as the variables of an equation and the actual-parameters as the values to be inserted in the equation. The calling statement:

   CompDiv(3,4,2,−1,r1,ig);

divides (3 + j4) by (2 − j) and the real part of the result is assigned to r1 and the imaginary part to ig. The variable actual parameters r1 and ig are used to return the results of this complex division to the main program. Make sure you do not confuse the concepts of value and variable parameters:

Value parameters: Conceptually, value-parameters provide the input values to a procedure. A value-parameter is declared in the procedure-heading; a value-parameter is local to a procedure and therefore is allocated computer memory space only while the procedure is being executed; this memory space is reclaimed by the computer after completion of the procedure execution. A value-parameter is initialised when the procedure is called and is assigned the value of the given actual parameters. There is no further interaction between the actual parameters and the formal parameters. An assignment within the procedure to a value-parameter has no effect on the corresponding actual-parameter. Therefore value-parameters cannot be used to return any results obtained within the procedure.

Variable parameters: Conceptually, variable parameters provide output from a procedure. Variable parameters are declared in the procedure-heading; their declaration must be preceded by the reserved word **var**. No new memory space is created during the execution of the procedure, but instead a memory space that was known previously by an identifier (usually global) is now also known by another name: the variable parameter name. A variable parameter is local in terms of name precedence (see rule 4 of identifier scope) but is global in effect. An assignment to a variable parameter is just like a direct assignment to the global variable it represents. Therefore variable parameters are used to return to the main program any results obtained from within the procedure.

Function **do not** have variable parameters in their formal parameter list.

Worked Example 5.7

The motion of a single pendulum, shown below, may be expressed by the first order differential equation:

$$\frac{d\omega}{d\phi} = \frac{-g}{L} \frac{\sin\phi}{\omega}$$

where ω = angular velocity in radians/s
ϕ = angle
g = 9.81 metres/s^2
L = length of pendulum

Given $L = 1$ metre, and as initial conditions $\phi_0 = 79°$, $\omega_0 = 0.8$ rad/s, determine the angular velocity as a function of angle.

This problem has a closed form solution:

$$\omega_0 = \sqrt{\left[\frac{2g}{L}\right](\cos\phi - \cos\phi_0) + \omega_0^2}$$

and can also be solved using numerical methods. The numerical method chosen here is due to Euler.

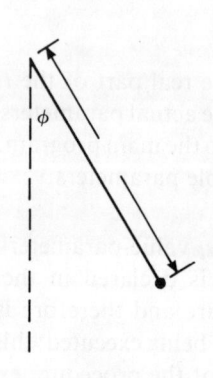

```
program pendulum(input, output);

(* Numerical solution to the pendulum problem *)

const degtorad = 0.01745329; g = 9.81; length = 1.0;

var startangle, startvelocity, anglestep : real;
    totalsteps : integer;

function value(angle, velocity:real):real;
begin
  value := g * sin(angle*degtorad)/(length*velocity)
end; (* function value *)

function exactsolution(angle, startangle, startvelocity:real):real;
begin
  exactsolution :=
    sqrt((2.0 * g / length) *
         (cos(angle * degtorad) - cos(startangle * degtorad))
       + startvelocity * startvelocity);
end;  (* function exactsolution *)

procedure differential(t0, y0, timestep:real; totalsteps:integer);

(* Solution to a first-order differential equation using
    Euler's method                                         *)

var time,y,k0 : real;
    step : integer;

begin
  writeln(t0:6:2, y0:12:6, y0:12:6);
  time := t0;
  y := y0;
  for step := 1 to totalsteps do
  begin
    k0 := value(time, y);
    y := y + k0 * timestep;
    time := time + timestep;
    writeln(time:6:2, y:12:6, exactsolution(time, t0, y0):12:6);
  end;
end; (* procedure differential *)

begin  (* main program *)
  write('Enter initial angle and angular velocity : ');
  readln(startangle, startvelocity);
  write('Enter anglestep and number of steps: ');
  readln(anglestep, totalsteps);
  writeln;writeln;
  writeln('  angle(rad)   velocity   exactsolution');
  differential(startangle, startvelocity, anglestep, totalsteps);
end.

Execution begins...

Enter initial angle and angular velocity : 10 1
Enter anglestep and number of steps: .2 25

  angle(rad)   velocity   exactsolution
  10.00    1.000000    1.000000
  10.20    1.340698    0.993977
  10.40    1.599846    0.987798
  10.60    1.821229    0.981460
  10.80    2.019399    0.974960
  11.00    2.201454    0.968296
  11.20    2.371509    0.961463
  11.40    2.532203    0.954459
  11.60    2.685352    0.947279
  11.80    2.832265    0.939920
  12.00    2.973926    0.932378
  12.20    3.111093    0.924647
  12.40    3.244364    0.916724
  12.60    3.374223    0.908603
  12.80    3.501066    0.900279
  13.00    3.625222    0.891747
```

```
         13.20    3.746967   0.883001
         13.40    3.866537   0.874034
         13.60    3.984133   0.864839
         13.80    4.099930   0.855410
         14.00    4.214079   0.845739
         14.20    4.326713   0.835817
         14.40    4.437951   0.825635
         14.60    4.547896   0.815185
         14.80    4.656640   0.804454
         15.00    4.764268   0.793433
      Execution terminated.
```

Worked Example 5.8 Solve the problem given in Example 5.7 using the Runge-Kutta method. The procedure given below performs the Runge-Kutta algorithm. Substitute this procedure in the previous example and comment on the results.

```
procedure differential(t0,y0,timestep:real; totalsteps:integer);

(* Solution to a first order differential equation using
   the Runge-Kutta Method                                    *)

var time,y,k0,k1,k2,k3,halfstep,change : real;
    step : integer;

begin
  writeln(t0:6:2, y0:12:6, y0:12:6);
  time := t0;
  y := y0;
  halfstep := timestep/2.0;

  for step := 1 to totalsteps do
  begin
    k0 := value(time, y);
    k1 := value(time+halfstep, y+k0*halfstep);
    k2 := value(time+halfstep, y+k1*halfstep);
    k3 := value(time+halfstep, y+k2*halfstep);
    change := (k0 + 2.0*(k1 + k2) + k3) * timestep / 6.0
    y := y + change;
    time := time + timestep;
    writeln(time:6:2, y:12:6, exactsolution(time, t0, y0):12:6);
  end;
end; (* procedure differential with Runge-Kutta *)
```

Worked Example 5.9 Write procedures to add, multiply and subtract two square matrices:

Solution:

```
program Matrices(input,output);

(* This program implements matrix arithmetic *)

const index = 3;
type twodarray = array  [1..index,1..index] of real;
    subscript = 1 .. index;
var  A,B,C : twodarray;

procedure AddMat(matA,matB:twodarray;var matC:twodarray);
var i,j:subscript;
begin
 for i:= 1 to index do
   for j:=1 to  index do
     matC[i,j]:=matA[i,j]+matB[i,j]
end (* AddMat *);

procedure SubMat(matA,matB:twodarray;var matC:twodarray);
var i,j : subscript;
```

```
begin
  for i:= 1 to index do
    for j:= 1 to index do
      matC[i,j]:=matA[i,j]-matB[i,j]
end (* SubMat *);

procedure MulMat(matA,matB:twodarray;var matC:twodarray);
var i,j,k: subscript;
    sum:real;
begin
  for i:= 1 to index do
    for j:= 1 to index do
      begin
        sum:=0.0;
        for k:= 1 to index do
          sum:=sum+matA[i,k]*matB[k,j];
          matC[i,j]:=sum
      end
end (* MulMat *);

procedure readMat(var matA,matB:twodarray);
var i,j : subscript;

procedure inputMatrix(var matrix:twodarray);
begin
  for i:= 1 to index do
    begin
      writeln('Enter elements of ',i:1,' row');
      for j:=1 to index do
        read(matrix[i,j]);
      readln
    end
end (* inputMatrix *);

begin
  writeln('Enter elements of first matrix ');
  inputMatrix(matA);
  writeln('Enter elements of second matrix');
  inputMatrix(matB)
end;

procedure writeMat(matC:twodarray);
var i,j : subscript;
begin
  writeln;
  for i:=1 to index do
    begin
      writeln(i:1,' row ');
      for j:=1 to index do
        write(matC[i,j]:5:2);
      writeln
    end;
  writeln
end (* writeMat *);

begin
  readMat(A,B);
  MulMat(A,B,C);
  writeln('Results of matrix multiplication : ');
  writeMat(C);
  AddMat(A,B,C);
  writeln('Results of matrix addition : ');
  SubMat(A,B,C);
  writeln('Results of matrix subtraction : ');
  writeMat(C)
end.

Execution begins...

Enter elements of first matrix
Enter elements of 1 row
1 1 1
Enter elements of 2 row
2 2 2
Enter elements of 3 row
1 1 1
```

```
Enter elements of second matrix
Enter elements of 1 row
2 2 2
Enter elements of 2 row
1 1 1
Enter elements of 3 row
2 2 2
Results of matrix multiplication :

1 row
 5.00  5.00  5.00
2 row
10.00 10.00 10.00
3 row
 5.00  5.00  5.00

Results of matrix addition :
Results of matrix subtraction :

1 row
-1.00 -1.00 -1.00
2 row
 1.00  1.00  1.00
3 row
-1.00 -1.00 -1.00

Execution terminated.
```

Procedural and Functional Parameters

So far functions which have value-parameters only have been discussed. No variable parameters are used since a unique result calculated by the function is returned by the function-identifier. Procedures which can have both value and variable parameters also have been considered. It is possible for these subprograms to have other functions/procedures as part of the formal-parameter list. These are called 'procedural and functional' parameters and are of value parameter type *i.e.* they can only input information to the subprogram. For example the function-heading:

function Integrate (upper,lower:*real*; **function** F(x:*real*): real):*real*;

defines a function named Integrate with formal parameters: upper, lower and a **function** F which produces a result of type *real* given a *real* parameter x. When the function is called the name of the actual function must be supplied together with all the other actual parameters. The actual function and the formal function must have the same number and type of parameters and must also return the same type of result; these functions are known as 'congruous' functions.

Programs that use procedural and functional parameters are often difficult to understand and even more difficult to debug. It is useful, at this stage, first to gain some experience with straight forward procedures and functions before attempting to use procedural and functional parameters.

For more details, see Welsh, J. and Elder, J. *Introduction to Pascal* (Prentice-Hall, 1982).

Worked Example 5.10 Write a Pascal program to evaluate:

$$e^x = 1 + x + \frac{x^2}{2!} + \frac{x^3}{3!} + \ldots\ldots$$

Solution:

This program is written using three **functions:** raise, factorial and sum. **function** sum uses the other two functions to calculate the series. One function is given as a functional parameter (factorial) and the other as an external function (raise). Both functions could be used as functional parameters or as external functions. The program is written in this form to highlight the the two possibilities.

```
program ExponentialSeries(input,output);

(* Generates series expansions of exp(x) *)

type posinteger = 0 .. maxint;
var exponent, result:real;
test:char;
terms : posinteger;

function raise(a:real;i:posinteger):real;
begin
 raise:=exp(i*ln(a))
end (* raise *);

function factorial(number:posinteger):posinteger;
var f,i: posinteger;
begin
 f:=1;
 if number=0 then factorial:=1
 else
   begin
     for i:=number downto 1 do
       f:=f*i;
     factorial:=f
   end
end (* factorial *);

function sum(function fact(i:posinteger):posinteger; x:real;
             limit:posinteger):real;
var subtotal:real;
 i:posinteger;
begin
  subtotal:=0.0;
  for i:=0 to limit do
    subtotal:=subtotal+raise(x,i)/fact(i);
  sum:=subtotal
  end (* sum *);

begin
  repeat (* y *)
    writeln;
    write('Enter exponent =');readln(exponent);
    write('Enter number of terms =');readln(terms);
    result:=sum(factorial,exponent,terms);
    writeln('exp(',exponent:1:2, ') = ',result:1:9,' using ',
                        terms:1,' terms');
    write('Type y to repeat calculations = ');readln(test)
  until test <> 'y'
end.

Execution begins...

Enter exponent =1
Enter number of terms =11
exp(1.00) = 2.718281826 using 11 terms
Type y to repeat calculations = n

Execution terminated.
```

Recursion

A subprogram may call other functions or procedures as seen in the previous sections. It is also possible for a subprogram to call itself. This condition is called 'recursion' and the subprograms using this method are referred to as 'recursive'. There are many mathematical expressions such as x^n and n! and polynomials, such as orthogonal polynomials, which are very useful in circuit theory and telecommunication theory, that can be evaluated using recursive subprograms. Recursion in a subprogram is indicated by the appearance of the subprogram's name as part of the statement part for procedures and as part of an expression for the case of functions. For example:

   factorial (number): = number * factorial(number−1);

The definition appears to be circular but it is not, since the arguments on both sides of the assignment statement are different. Although defining something in terms of itself may look strange, it is seen that the definition is a perfectly sensible one. Pascal is one of the few languages that permits recursion. To compute the value of a function the definition is used repeatedly. The repetition terminates when the stage is reached when the function-identifier whose value is being calculated can be found without making use of the function being defined. An *infinite* recursion occurs if this condition cannot be met. An example helps to clarify these points:

```
function factorial(number:integer):integer;
begin
  if number = 1 then factorial: = 1
                else factorial: = number * factorial(number−1)
end (* factorial *);
```

Suppose it is required to compute factorial(4). According to the recursive definition given above:

       factorial(4) = 4 * factorial(3)
  where factorial(3) = 3 * factorial(2)
  where factorial(2) = 2 * factorial(1)
  and   factorial(1) = 1

Now factorial(1) is not defined in terms of the function to be defined, therefore this is our exit condition. As can be seen, the above example requires more memory storage than the equivalent repetitive function discussed earlier. The choice between iteration and recursion is usually determined by (i) the need for temporary storage and (ii) elegance of solution.

Worked Example 5.11 Write a recursive and a non-recursive procedure to reverse the digits of a given four-integer decimal number.

Solution:

```
program Reverse(input,output);
(* reverses digits in a number *)
type posinteger = 0 .. 9999;
```

```
    var decimal : posinteger;

procedure Revrs1(number:posinteger);
var thousands,hundreds,tens,ones,reverse:posinteger;
begin
 thousands:=   number div 1000;
 hundreds :=  (number mod 1000) div 100;
 tens     :=  (number mod 100) div 10;
 ones     :=   number mod 10;
 reverse  :=  (1000*ones)+(100*tens)+(10*hundreds)+thousands;
 write(reverse:1)
end (* Revrs2 *);

begin
  write('Enter positive integer number =');readln(decimal);
  write('Using non-recursive procedure =');Revrs1(decimal);
  writeln;
  write('Using recursive procedure ='); Revrs1(decimal);
  writeln
end.

Execution begins...

Enter positive integer number =1234
Using non-recursive procedure =4321
Using recursive procedure =4321

Execution terminated.
```

It can be seen from the above that the recursive procedure can cope with integers larger than four digits.

Forward Directive

Large programs can contain a very large number of subprograms and the usual practice is to write these in alphabetical sequence so that they can be found easily by the reader. This sometimes leads to a problem: a subprogram could be called in another subprogram before it was defined. This problem can be overcome in Pascal by the use of forward directive. Pascal allows the definitions of a subprogram's heading to be in a different part of the program from the subprogram's body as long as both are within the same block. The subprogram-heading must precede the definition of the body. For example:

```
      ................
      function tan(x:real):real;
        forward;
      ................
      other procedures and functions
      ................
      function tan;
      begin
        ................
      end (*tan*);
      ................
      ................
```

Note that unless the function body was defined the formal parameter list was not included. If it is, an error is flagged. If it is required to use forward subprogram reference it is strongly recommended that you follow the common practice of repeating the formal-parameter list when the subprogram's body is defined but include the list in a comment *i.e.*

>   **function** tan (* (x:*real*):*real* *);
>   **begin**
>   ....
>   **end** (* tan *);

This improves program readability and helps in debugging the program.

Summary

Modularity is an important part of well-written programs. Programs which subdivide their tasks into smaller subprograms are easier to read, write and understand. They are also much simpler to correct and can be substituted by better versions without major rewrites. The best method of writing a program is to follow the top-down approach in which the problem is described in its most general form and then subdivided into smaller subprograms, always keeping in mind the objectives of the main problem.

Pascal provides two subprograms: **function** and **procedure**. These subprograms are very powerful tools for writing modular programs. All subprograms used in a Pascal program must be declared after the **var** declarations and before the main body of the program. The order of the subprogram definitions must be such that each subprogram is defined before it is used in another subprogram. In some cases subprograms must be declared out of sequence, this problem can be overcome by the use of the forward directive. When the subprogram is defined after a forward directive, the formal parameter list and type (for functions) is omitted but should be inserted as a comment.

Pascal uses the block structure convention. Syntactically a block consists of a program or subprogram heading, followed by definitions/declarations and a statement part. A Pascal program can be considered as the outer block with a number of smaller blocks inside it. An identifier in a Pascal program has scope. Reference is made to the scope of an identifier as the block in which the definition/declaration of that particular identifier is made, including any blocks enclosed by that block. There are certain rules which must be obeyed when considering the scope of identifiers.

Subprograms may contain definitions/declarations of **label, const, type, var** and other subprograms. These new parameters are considered as local to the subprogram and have no meaning to any other subprogram or the main program. They can be used only in the subprogram itself. Local variables exist in computer memory only while the subprogram is being executed; they are destroyed after completion of execution. To preserve information from one subprogram to the next it is necessary to use global variables. Identifiers which have been defined/declared in the main program are global identifiers and can be used anywhere in the program. If a local identifier has the same identifier-name as a global variable then the local variable takes precedence.

Parameters which are used within a subprogram can also be declared in the subprogram's heading as formal-parameters. Formal parameters can be of any Pascal type or even other subprograms. These parameters are place keepers for the calling parameters. Formal parameters can be value parameters for both subprograms. Only procedures may have variable parameters.

Communication between a subprogram and other subprograms and the main program is done by means of parameter passing. When a subprogram is called the formal parameters and the calling parameters, also called actual parameters, are matched in the order they are listed. These parameters must also correspond in type. Both procedures and functions accept as input parameters that have been declared as value parameters in the formal parameter list. Subprograms differ in the way they return their output. A function-identifier represents a unique value and its heading must specify the function's type of result. Also, the function's statement part must include an assignment statement that gives the function its value. Therefore for output from a **function** the function-identifier is used. Procedures are used if more than one output value is required or it is wished to perform certain tasks requiring no output. Output from a **procedure** is obtained by using the variable parameters declared in the formal parameter list.

An alternative form of communication with subprograms is to use the global variables. If this method of communication is used the modularity of the program is destroyed. A local variable assignment to a global variable is called a side-effect. Side-effects can be dangerous and should be used with extreme care.

Subprograms may include recursive calls of themselves. This programming technique can simplify immensely the implementation of iterative algorithms. The main drawback of this method is the amount of memory that is required for its execution.

Problems

5.1 Explain the difference between (i) procedure and function, (ii) local and global variable, (iii) value and variable parameter and (iv) actual and formal parameter.

5.2 Write subprograms to evaluate:
   (i) $\sinh^{-1} x = \ln(x + \sqrt{x^2 + 1})$
   (ii) $\cosh^{-1} x = \ln(x \pm \sqrt{x^2 - 1})$ (use plus sign if $x > 1$)
   (iii) $\tanh^{-1} x = \dfrac{1}{2} \ln\left(\dfrac{1+x}{1-x}\right)$ for $x^2 < 1$

5.3 Write a function which converts a given hexadecimal number to a decimal integer number.

5.4 Write a procedure to invert a given matrix

5.5 Evaluate the determinant of a square matrix

$$[Hint: \text{Det}(A) = \sum_{j=1}^{N} (-1)^{i+j} A_{ij}(\text{co-factor}_{ij}(A))]$$

5.6 Using recursion write a single procedure to convert a given decimal number to an octal or binary number.

5.7 Write recursive and non recursive functions to calculate the coefficient of Chebyshev polynomials given that

Horrocks, O.H. *Feedback Circuits and Op.Amps* (Van Nostrand Reinhold, 1983).

DiStefano, J.J., Stubberub, A.R. and Williams, I.J. *Feedback and Control Systems* (McGraw Hill, 1967).

$$C_0(\omega) = 1$$
$$C_1(\omega) = \omega$$
$$C_{n+1}(\omega) = 2\omega C_n(\omega) - C_{n-1}(\omega)$$

and compare the two functions.

5.8 Given the open loop transfer function of a network plot the Nyquist diagram.

5.9 Incorporate in your program for Problem 5.8 the calculation of gain and phase margin.

5.10 The trapezoidal rule for numerical integration is given by

$$\int_a^b f(x)dx \cong h\left[\tfrac{1}{2}f(a) + f(x_1) + f(x_2) + \ldots + f(x_{n-1}) + \tfrac{1}{2}f(b)\right]$$

write a subprogram for this algorithm and hence evaluate

$$\int_0^2 (x^3 - 4x + 4)dx \qquad \text{with } h = 0.5$$

Structured Data Types: Array, File, Set and Record. The Pointer Data Type

6

Objectives

☐ To introduce the various Pascal structured data types.
☐ To introduce the **array** data structure.
☐ To introduce the concepts of **packed** and unpacked **array**.
☐ To use the **file** structure.
☐ To introduce the **set** and **record** data structures.
☐ To mention the pointer data type.

In Chapter 1 the Pascal data types were classified into three groups: simple, pointer and structured data types. A structured data type is one whose values are each made up of simpler component values. So far the simple data types and the structured data type **array** have been considered. In this chapter the pointer data type and the various Pascal structured data types are examined. These topics are not covered in great detail; only the parts relevant to first year electronic engineering students are emphasised. Many concepts introduced in this chapter are used in Chapter 7.

The Array Structure

The array is the simplest of all available structures in Pascal. Its data structure is a contiguous fixed length sequence of data objects, all of the same type. An array definition has the form:

    type array-type = **array** [subscript-type] **of** element-type;
    var identifier : array-type;

or, using short-hand notation:

    var identifier : **array** [subscript-type] **of** element-type;

The arrays encountered so far have subscripts of type *integer* and elements of type *integer* or *real*. In fact, the most commonly used subscript is of type *integer*. Arrays can have subscripts of any ordinal type and elements of any simple or structured data type. Examples of **array** with elements of simple data type are:

    const n = 10;   m = 15;
    type initl = 'a' .. 'r';
           upper = 'A' .. 'Z';
           RGB   = (Red,Green,Blue);
           colour = **array** [RGB] **of** *real*;

The elements of an array can be also referred to as components of an array and the subscript as the index.

Parentheses in Pascal are used with subprogram and enumerated type definitions, not with arrays.

107

var TVspot : colour;
    test : **array** [letter] **of** *boolean*;
    vector : **array** [−n..n] **of** *real*;
    stack : **array** [0..m] **of** *integer*;
    shift : **array** [1..26] **of** upper;
    line : **array** [1..80] **of** *char*;

As mentioned above arrays can have any structured data type as element-type including arrays, for example:

var oneD : **array** [1..5] **of** *real*;
    twoD : **array** [1..10] **of array** [1..5] **of** *real*;
    threeD: **array** [1..15] **of array** [1..10] **of array** [1..5] **of** *real*;

The last two declarations can be simplified to:

var twoD : **array** [1..10, 1..5] **of** *real*;
    threeD: **array** [1..15, 1..10, 1..5] **of** *real*;

Subscripts of any ordinal type could have been used for the above examples. There are no limits placed on the dimensions which an array can have, but beware of storage limitations! Arrays with one subscript are known as 'one-dimensional' or 'vector' arrays; arrays with two subscripts are called 'two-dimensional' or 'matrix' arrays; arrays with more than two subscripts are called multidimensional arrays.

An array with subscripts [1..100,1..100,1..100] has 1 million elements!

If two arrays have identical declarations *i.e.*

var A,B:**array** [1..10] **of** *integer*;

then one array can be copied to the other using the assignment statement:

A := B;

This is equivalent to:

for i: = 1 **to** 10 **do**
    A[i] := B[i];

Remember that, the subscript type does not have to be an integer, neither does the control variable of a for-statement.

Arrays and for-statements are very well suited to each other. The for-statement steps the value of the subscript through a series of values, thus allowing the corresponding array elements to be referenced.

Before proceeding any further let us summarise:
(i) An array is a fixed sequence of objects that can be denoted by a single identifier.
(ii) An array declaration specifies the range of subscripts that an array can have and the type of its elements.
(iii) The elements of the array can be defined as simple or structured data type.
(iv) The type of subscript must be of ordinal type.
(v) Any element from the array sequence can be obtained by specifying the array identifier followed, in square brackets, by the subscript value.
(vi) The subscript value indicates the position of the element in the array.
(vii) Arrays can be accessed randomly.
(viii) In a program an array subscript can be written as a variable or as an expression, provided its type is compatible with the array subscript-type.
(ix) An array must have a fixed number of elements.
(x) Arrays can be multidimensional.

Arrays as Subprogram Parameters

As discussed in Chapter 5, an array can be used as an actual parameter to transfer data to a subprogram if it is declared as a value parameter. An array may also return data to the calling program if it is declared as a variable parameter. Arrays cannot be used to return values from a function since a function returns a single-value result. Before an array can be used as an actual parameter it must be previously declared in the block enclosing the subprogram. Remember the type of the parameter in the formal parameter list can only be specified as an identifier.

As specified in standard Pascal (BS6192).

Value parameters have their values copied, for use in a subprogram, in new memory locations. Within a subprogram all references to value parameters are references to the copy of the parameters. Transferring an array as a value parameter involves copying the entire array. For large arrays this can be wasteful in both processor time and memory space. Program efficiency can be improved by transferring arrays to a subprogram as variable parameters rather than as value parameters. By making the array a variable parameter, the subprogram gains access to the array already stored in memory thus eliminating the need for copying. For example if we want to take advantage of this method when the **procedure** plot was written (Example 5.4), then the procedure heading would have been written as:

    procedure plot(**var** x:onedarray; **var** y:twodarray; Points,Plots:posinteger);

An array used as an actual parameter must be identical in type to the corresponding formal parameter array *i.e.* it must have the same bounds, must have the same subscript type and also the same element type. Also remember that subprogram parameters must have their type specified by an identifier (see Example 5.4). Some occasions arise when it is required to use the same subprogram a number of times but with arrays having the same specifications except for different number of elements. This problem can be overcome with the use of 'conformant array parameters', a method which is error-prone and, since it is outside the scope of this book, is not discussed further. The reader is referred to the first two books in the bibliography for further information.

A word of warning: beware of side-effects!

For more details see: Welsh, J. and Elder, J. Introduction to a Pascal (Prentice-Hall, 1982).

Write two subprograms to evaluate the dot and cross product of two vectors.

Worked Example 6.1

Solution: The following definitions are assumed:

    type   nsub = 1..10;
               msub = 1..15;

    nvector = **array** [nsub] **of** *real*;
    mvector = **array** [msub] **of** *real*;
    matrix   = **array** [nsub, msub] **of** *real*;

(i) The evaluation of the dot product can be written as a function since a single scalar value results from the evaluation of the dot product of two vectors.

    function DotProd(**var** p,q:nvector; n:nsub):*real*;
    var i:nsub;
        dp : *real*;

```
begin
    dp: = 0.0;
    for i: = 1 to n do
        dp: = dp + p[i] * q[i];
    DotProd: = dp
end (* DotProd *);
```

(ii) The evaluation of the cross product must be written as a procedure since the required result is a matrix:

```
procedure CrossProd(var p:nvector; n:nsub; var q:mvector; m:msub;
                    var r:matrix);
var i:nsub;
    j:msub;
begin
    for i: = j to n do
        for j: = 1 to m do
            r[i,j]: = p[i] * q[j]
end (* CrossProd *);
```

In both solutions the arrays were declared as variable parameters. If memory space is not at a premium then the arrays could be declared as value parameters, for example:

(i) **function** DotProd(p,q:nvector;n:nsub):*real*;

(ii) **procedure** CrossProd(p:nvector;n:nsub;q:mvector;m:msub;**var** r:matrix);

The author's preference is to declare arrays in this way for the extra security that is provided by declaring parameters as value parameters.

Packed Arrays

An array may have a very large number of elements and therefore requires a large amount of memory space. In Pascal, the reserved word **array** may be preceded by another reserved word **packed** which tells the compiler to pack the array elements as efficiently as possible. This usually indicates that more than one value is stored in each memory location, which means that the time taken to access these values is increased. A **packed array** can be used like any other **array** within the program. A typical definition is:

```
type pline : packed array [1..80] of char;
     line  : array [1..80] of char;
```

The effectiveness of the packing depends on the type of the array elements. It is efficient if the array elements are *boolean, char,* enumerated or subrange type. Packing an array has no effect on the results of the program, it only reduces memory space and increases the time required to read or write into memory. There are a number of points which should be considered before packing an array:

(i) An element of a packed array cannot be transferred as a variable parameter into a subprogram.

(ii) If a program frequently uses elements of a packed array, the speed of execution is severely affected.

(iii) Packing an array with *real* and *integer* type elements has very little effect on memory space.
(iv) A good compiler would minimise storage space itself, without a need for packing by the programmer.

If the elements of a packed array are referenced frequently in a program, we can improve the time penalty by unpacking and then packing the array before and after the need arises. Pascal provides two procedures to do this: *unpack* and *pack*.

Assuming the above array definitions, a procedure call *pack*(line,i,pline) copies the elements from the unpacked array named line, starting with element line[i], to the packed array pline, starting with the first element pline[1]. A procedure call *unpack*(pline,line,i) has the opposite effect. Note the difference in the position of the parameters between these two procedures and that the size of the **packed array** determines the number of values which can be packed or unpacked with these two procedures.

Strings

A sequence of characters is known in Pascal as a 'string'. A string constant consists of a sequence of characters enclosed in single quotes *i.e.*

    const memo = 'This is a string constant';
           code = 'A645';

Character strings have been used before with the two output procedures:

    write('This is a string constant');   which is equivalent to *write*(memo);
    writeln('A645');   which is equivalent to *writeln*(code);

memo and code are considered in Pascal to be constants of **packed arrays** *i.e.*

    memo is a constant of type **packed array**[1..25] **of** *char*
                           (* memo has 25 characters *)
    code is a constant of type **packed array**[1..4] **of** *char*
                           (* code has 4 characters *)

packed arrays of type *char* are the only array type for which *array* constants are provided. For all other type of arrays, constant values must be assigned component by component using a for-statement.

Therefore, a sequence of **packed array** elements of type *char* is known as a string type. Even though string is not declared as a standard data type, it has been accepted practice to call a variable of string-type a variable which is of type **packed array** with *char* elements and *integer* subscripts with the following two restrictions:
(i) The elements must be of *char* type and not a subrange of *char*.
(ii) The subscript is a subrange of *integer* and not the whole *integer* range. The lower bound of the subrange is 1 and the upper bound is greater than 1.

No variables are allowed in the definition of array bounds.

For example:

    type string = **packed array** [1..9] **of** *char*;
    var elem1,elem2 : string;

In Pascal fixed length character strings have the following properties:
(i) All six relational operators can be used to compare the lexicographic ordering of two equal length strings. This ordering depends on the available character set *e.g.*

Strings with variable length with the same definition are not allowed.

    elem1 > elem2   elem1 < > elem2
    elem1 = 'character'   elem2 < ' test '

Note elem1 and elem2 have no subscripts.

(ii) An entire string can be output all at once by specifying one variable rather than output one character at a time.

    write(elem1);   *writeln*(elem2, ' ',elem1);

(iii) We can use the assignment operator with string variables.

    elem1 := elem2;   elem2:= 'Inductor ';
    elem1 := '   opAmp';   elem2:= ' diode ';

(iv) Arithmetic operations cannot be used with strings.

(v) The procedure *read* and *readln* cannot be used to input strings as one entity. Strings are input into the program a character at a time:

```
for i:= 1 to 9 do
  begin
    read(ch);
    elem1[i] := ch
  end;
readln;      (* takes care of the carriage return *)
```

Worked Example 6.2   Write a Pascal program to sort a given list of items in lexicographic order.

Solution:

```
program Order(input,output);

(* program sorts a given list of items in lexicographic
   order.  the Quick Sort method has been used           *)

const listlength = 15;
      stringlength = 10;
type  position = 0 .. listlength;
      string = packed array [ 1.. stringlength] of char;
      itemlist = array [position] of string;
var   sequence : itemlist;
      counter,i : position;

procedure readlist (var list:itemlist; var number:position);
var entry : string;

procedure readstring(var item:string);
var ch : char;
    length : 0..stringlength;
begin
  length :=0;
  (* items are entered one per line *)
  while not eoln and (length <= stringlength) do
  begin
    read(ch);
    length := length+1;
    item[length] := ch
  end;
  readln;
  (* add extra spaces to make strings equal length *)
  for length:=(length+1) to stringlength do
    item[length] := ' '
end (* readstring *);

begin
  number := 0;
  while not eof do
    begin
      number := number+1;
```

```
            readstring(entry);
            list[number] := entry
         end
end (* readlist *);

procedure QS(var list:itemlist; first,last:position);
(* procedure to sort list *)
var partition,temp:string;
    low,high,middle : position;
begin
   (* set upper, lower and middle limits *)
   low := first;  high := last;   middle := (first+last) div 2;
   partition:=list[middle];
   repeat
      while list[low] < partition do
          low := low+1;
      while list[high] > partition do
          high := high-1;
      if low <= high then
        begin
           (* exchange *)
           temp := list[low];
           list[low] := list[high];
           list[high] := temp;
           low :=low+1;
           high :=high-1
        end;
   until (low > high );
   if first < high then QS(list,first,high);
   if low < last then QS(list,low,last)
end (* QS *);

begin
   writeln('Enter items one per line.');
   writeln('Terminate list with end-of-file.');
   readlist(sequence,counter);
   QS(sequence,1,counter);
   writeln('The sorted list has ',counter:1,' items :');
   for i:=1 to counter do
      writeln(sequence[i])
end.

Execution begins...

Enter items one per line.
Terminate list with end-of-file.
fred
Overnight
dragon
lot123

The sorted list has 4 items :
Overnight
dragon
fred
lot123

Execution terminated.
```

If the list is very long and sorting takes a long time try unpacking, sorting and then packing your array.

The File Structure

In Pascal a stream of data items stored sequentially in auxiliary memory is known as a 'file'. Pascal uses only 'sequential' files, the data items or 'components' of which can be accessed only in the order in which they are stored *i.e.* no random access is permitted. All component values within a file must be of the same type- *e.g.* any simple or structured type provided they are not, nor include, a file-type.

In Pascal the syntax description of a file is:

Two files that Pascal provides: *input* and *output* have already been met.

    type file-type = **file of** component-type;
    var identifier : file-type;

or using the shorthand notation:

    var identifier : **file of** component-type;

For example:

    type intfile = **file of** *integer*;
          relfile = **file of** *real*;
          chrfile = **file of** *char*;
    var Ifile : intfile; (∗ file contains only integer components ∗)
         Rfile : relfile; (∗ file contains only real components ∗)
         Cfile : chrfile; (∗ file contains only character components ∗)

> The original report allowed only textfiles and record files. There are three main categories for files: sequential, direct-access and indexed files.

Files in Pascal are created for two reasons:

(i) To use the data items after the completion of a program, in which case store these data items in the computer auxiliary memory. These files are called 'external' files (permanent) and exist after the execution of the program.

> Like an array all file components must be of the same type. But unlike an array the size is not specified.

(ii) To use data items created in one part of the program for another part. It is not necessary to store these data items in peripheral memory since the transfer is done during program execution. These files are called 'internal' files (temporary) and exist only for the duration of program execution.

External files need to be accessed by other programs and therefore must be specified in the program heading:

    program Test(*input,output,*InFile,OutFile);

> **program** heading parameters establish communication links.

Every external file created by the programmer must be declared as a variable in the main program except the standard files *input* and *output* which are automatically declared by Pascal. Internal files can be considered as analogous to local variables in a subprogram. Internal files should be declared in the block where they are being used.

> The file grows as new items are added to the end of the file.

The entire contents of a sequential file are not accessible all at once nor can we jump into the middle of a file. A file can be written into sequentially by adding new data items one at a time at the end of the file. Components of sequential files can be read only in the order in which they were written, always starting at the beginning.

All Pascal files can be read and written sequentially from the beginning, but we cannot simultaneously read and write the same file. Therefore, there is no need for the complete file to be stored in the computer main memory, only the current component needs to be stored there.

The declaration of a file automatically creates a window into that file. This window allows the value of one component to be seen at a time (the one in main memory), the other values (in auxiliary memory) are not accessible. For example, assume that the Ifile declared above has 6 components:

    3   5 |7| 11   13   17

The window, shown above, allows the third component to be seen in the file. The window is usually referred to as the 'file-buffer' and is indicated by the special up-arrow (↑) or circumflex (^) symbol. For example the **var** declarations above have also created automatically three file-buffers: Ifile^, Rfile^ and Cfile^.

Standard Pascal Procedures for Files

Any file and its file-buffer can be used with four standard Pascal procedures and with the boolean standard identifier function met before: *eof*. The following file declaration is used in describing this section:

    var F : **file of** *real*;   (∗ short-hand notation ∗)
          x : *real*;

Neither the assignment operator nor the relational operators can be used with file types since only one component at any one time is stored in the computer's main memory.

To insert data values into a file first call the procedure *rewrite*.

(1) *rewrite*(F): This procedure call initialises a file for writing. Any data previously in the file is destroyed. The file-buffer F^ is set to the first component of the file ready for writing. All output files, except *output*, must be initialised before they are used.

A component can be written into the file by assigning its value to the file-buffer and then calling the procedure *put* i.e.

    F^: = x;   *put*(F);

(2) *put*(F): This procedure call writes the contents of F^ onto the end of file F.

The above pair of statements is repeated for subsequent additions to the file, always adding the new values at the end of the file. Owing to the frequent use of these two statements they have been combined into the standard procedure *write* i.e.

    write(F,x);

which means write the value of x into file F. If the file is not specified then the file *output* is assumed.

To read values from an existing file the procedure *reset* should be called.

(3) *reset*(F): This procedure call has the effect of positioning the file-buffer F^ to the first component of the file F ready for reading. All files, except *input*, must be reset before reading.

The value of a component can be read from the file by assigning its value from F^ and then moving the file-buffer to the next component of the file by calling the standard procedure *get* i.e.

    x: = F^;   *get*(F);

(4) *get*(F): This procedure call advances the file-buffer F^ to the next component i.e. it assigns the value of this component of the file F to the file buffer F^.

Again the previous two statements can be replaced with the *read* standard procedure: *read*(F,x). As in the case with *write*, if the file argument is omitted then the file *input* is assumed.

In the above examples only one variable was read or stored. It is equally correct to read or store a number of values using the same procedure statement provided all values are of the same type as specified in the file declaration i.e.

    read(F,a,b,c,d);   (∗ reads the values of a,b,c and d from file F ∗)
    write(F,a,b,c,d);   (∗ writes the values of a,b,c and d into file F ∗)

Another useful function available in Pascal for use with files is *eof*.

> We met *eof* in Chapter 3.

(5) *eof*(F): This function call returns *true* if the file's file-buffer has moved beyond the end of the file F, otherwise it returns *false*.
The file-buffer is undefined when *eof*(F) is *true*.
It is an error to try to inspect it.

The following points should be noted:
(i) *eof*(F) is set to *false* after a *reset*(F) if the file is not empty.
(ii) *eof*(F) is set to *true* after a *rewrite*(F).
(iii) *eof*(F) remains *false* after a *get*(F) unless the last value has been read in. In this case *eof*(F) is set to *true*.
(iv) *eof* remains *true* after a *put*(F)
(v) *eof* without a file parameter is assumed to apply to the standard file *input*.

Summarising what has been learned so far regarding files:
(i) Pascal can operate only on sequential files.
(ii) All file components within a file must be of the same type.
(iii) File size is not specified; files are allowed to expand and contract as necessary.
(iv) Two kinds of file are available: external and internal.

> Implementation dependent.

(v) External files must be specified in the program heading.
(vi) Both internal and external files must be declared to be of file-type.
(vii) When a file is created a file-buffer is automatically available.
(viii) A number of standard Pascal subprograms are available for use with files.

Worked Example 6.3 (1) Write the necessary code to find the mean value of a number of *integer* numbers stored in a file.

(2) Write the necessary code to store all *integer* values greater than 100 read from file F1 into file F2.

Solution:

> It is quite common to forget to *reset* or *rewrite* files before using them!

1. ```
   sum: = 0; i: = 0;
   reset(F);
   while not eof(F) do
     begin
       sum: = sum + F^;
       i: = i + 1;
       get(F)
     end;
   mean: = sum/i;
   ```

2. ```
   reset(F1); rewrite(F2);
   while (not eof(F1)) do
     begin
       while (F1^ > 100) do
         begin
           F2^: = F1^;
           put (F2)
         end;
       get (F1)
     end
   ```

Textfiles and Standard Procedures

If sequential files containing *char* components are divided up into lines, then they are called 'textfiles'. Each line is terminated by a special component value known as a 'line control character' or simply 'line-separator'. Pascal has a predefined file-type called *text* that recognises this division into lines. The standard files *input* and *output* are of type *text*. Textfiles can be defined in the **var** declaration part of the program as:

var F1,F2: *text*;

The only way to generate a line-separator is by calling the standard Pascal procedure *writeln i.e. writeln*(F1). The line-separator can be detected by using the boolean function *eoln*. The function call

eoln(F1);

returns the value *true* if the current character value from the textfile F1 is a line-separator. In this case the character value given by the file-buffer F1$^\wedge$ is blank *i.e.* the line-separator value appears as a space.

> The argument of *eoln* must be a textfile.

Textfiles can be interpreted as sequential files of characters or as a sequence of lines of characters.

All subprograms discussed in the previous section can be also applied to text-files. Remember that if a file is not specified when subprograms are called then the standard files *input* and *output* are assumed.

reset(*input*) and *rewrite*(*output*) are automatically performed by Pascal at the beginning of the program. If these two procedures are called again in the middle of the program, with the standard files as arguments, then it may cause an error (implementation dependent).

The availability of a file-buffer with the standard file *input* (*input*$^\wedge$) is very useful for inspecting a character before it is read into a program. For example, to count the number of 1s read in directly from *input* consisting of binary bits then:

```
noofones := 0;
while not eof do
   begin
     if input^ = '1' then      (* check for 1 *)
        noofones := noofones + 1;
     get(input)                (* move to next character *)
   end;
```

The standard procedures *read* and *write* can be used with textfiles other than *input* and *output*:

```
read(F1,a,b,c,d);
write(F2,a,b,c,d);
read(input,a,b,c);    (* is equivalent to read(a,b,c); *)
write(output,a,b,c);  (* is equivalent to write(a,b,c); *)
```

The standard procedure *readln* also can be used with textfiles. It is interesting to note that this procedure is defined as:

```
begin
   while not eoln(F1) and not eof(F1) do
       get(F1);
   get(F1)
end;
```

> It is a good programming practice to check for *eof* before using a file.

i.e. F1^ is advanced to the first character of the next line.

The standard procedure *writeln* is defined only for textfiles and not for any other type of sequential file. A procedure call *writeln*(F2) has the effect of adding a line-separator character to the file F2.

The standard procedure *page* may be used with textfiles. This procedure when called, *page*(F2), has the effect of moving the writing of the output to the top of a new page.

Worked Example 6.4 A marine echo sounder is used to obtain the topography of the bottom of a lake. The return signal is converted into a real number representing the distance between the echo-sounder and the bottom of the lake. A series of measurements is stored in a file. Before any further processing is done using these numbers, the data is *cleaned* from spurious noise by setting an upper and lower limit for the acceptable distance measurements. Another source of error could be due to fish passing under the boat while the experiment was carried out. To eliminate this source of error the cleaned signal is averaged over a number of samples. Write a Pascal program to perform this task.

Solution: In real life the amount of data collected, in say half an hour, fills a floppy disk but the program given below is equally applicable to the small file used here. To overcome the fish problem samples are usually averaged over at least 100 samples (depending on the sampling frequency) but for our test Nsamples is set to 5.

```pascal
program RemoveNoise(output,Data,Outfile);

(* Removes noise from a signal *)

const Nsamples = 5;
var Data : file of real;
    Outfile : text;
    mean,sum,value : real;
    counter,i : integer;
begin
  reset(Data);   rewrite(Outfile);
  sum :=0.0;   counter := 0;   i:= 0;
  while not eof(Data) do
   begin
     if (Data^ >= 2.0) and (Data^ <=50.0)
        then
          begin
            read(Data,value);
            sum:=sum+value;
            i:=i+11
          end
        else   get(Data);    (* move to next value *)
if i=Nsamples then
                begin
                  mean:=sum/Nsamples;
                  write(Outfile,mean:4:2);
                  counter:=counter+1;
                  sum:=0.0;
                  i:=0
                end
   end (* while loop *);
   write('The Outfile has ',counter:1,' components')
end.
```

A data file of type *real* was created before the program was run. Make sure that when you create a similar file you do so using a simple Pascal program to generate your numbers rather than using your editor. The file used for this example, Data, had the following sequence of real numbers:
1.00 2.00 3.00 4.00 5.00 5.00 10.00 15.00 20.00 25.00 30.00 35.00 20.00 40.00 60.00 80.00 100.00 1.00 4.00 9.00 16.00 25.00 36.00

If you use your editor you create a *text* file.

Typical Output : The Outfile has 3 components

The file Outfile when listed contained the following values: 3.80 20.00 21.60

The Set Structure

Pascal is one of the very few programming languages which allows a programmer to operate with sets. Sets are used when there is no concern with the order in which the elements are stored or with the number of times a value is repeated in a set.

A set is defined as a collection of distinct objects all of the same ordinal type. Any set may contain some, all, or none of the objects of that type. For example, the set [1, 5, 7, 9, 11] has five 'elements' or 'members' which are of 'base-type integer'. In set expressions, individual elements of a set are placed between square brackets. In Pascal a set structure is defined in the **type** definition part of the program. The definition uses a set-type, an equal sign followed by the reserved words **set of** and the base-type. The base-type must be of ordinal type. For example:

In mathematics the elements of a set are enclosed in braces. But since Pascal uses braces for comments, the square brackets are used for sets.

    type punctuation = **set of** *char*;
    var symbol : punctuation;

In the above definition, punctuation is the set-type which is associated with the ordinal type *char*. The values that the variable symbol can take are the whole available character set or a subset of the character set *i.e.*

    symbol : = [' , ' , ' . ' , ' ; ' , ' (' , ') ' , ' – '];

Consider a few more examples:

    type components   = (Inductor,Resistor,Capacitor,Diode,Transistor,
                                  OpAmp);
            circuit   = **set of** components;
            TVcolours   = (Red,Green,Blue);
            colour   = **set of** TVcolours;
    ....
    var circelem,passive,active : circuit;
           dot   : colour;
    ....

The shorthand notation for defining sets *e.g.* **var** dot: **set of** (Red,Green,Blue); could be used.

The above definitions and declarations permit:

119

> Remember that the two dots .. mean through and including.

circelem := [Inductor .. OpAmp];	(* a set with 6 elements *)
passive := [Inductor,Resistor,Capacitor];	(* a set with 3 elements *)
active := [Diode,Transistor,OpAmp];	(* a set with 3 elements *)
circelem := [Inductor..Diode,OpAmp];	(* a set with 5 elements *)
spot := [Red,Green,Blue];	(* a set with 3 elements *)

They also allow:

circelem := [Diode];	(* a set with 1 element *)
passive := circuit;	(* a set with 6 elements *)
active := [Diode,OpAmp];	(* a set with 2 elements *)
spot := [];	(* a set with 0 elements *)

Therefore, an element specification can be made by listing the assignment values between square brackets or by supplying the name of the set-type. The last assignment above 'spot := [];' is the 'empty set' i.e. a set which contains no elements. Every set includes the empty set as a value regardless of the base-type. The following points can be made regarding sets:

(i) Elements of a set must be of the same ordinal type.
(ii) If there are more than one element in a set, then the elements are separated by comma.
(iii) If a set consists of a contiguous range of values then the abbreviation [x..y] may be used.

> If x > y then [x..y] denotes the empty set.

(iv) A set may contain no elements. This is the empty set.
(v) The order in which elements are listed is not important nor is the number of times an element is repeated in a set.
(vi) Elements of a set may be constants, variables and/or expressions, as long as they all are of the same ordinal type *i.e.* [1, 7, 15 **div** 4] is acceptable.
(vii) Unlike an **array** we can access individual set elements as a single unit, without the need of an index or subscript.

> Think of the number of possible set assignments that are possible with type base *integer* (2^{maxint}). Use *integer* subrange to limit the size.

(viii) If c is the cardinality of the base-type then there are 2^c possible values that the set can have.

Point (viii) above states, for example, that the set circelem may have 2^6 possible set assignments. Although, the recently accepted BS6192 standard does not impose any restrictions on the base-type of sets, in fact a number of implementations place a limit on the cardinality of the base-type. This limit varies from implementation to implementation and is typically between 48 and 2048 elements. This curtails the portability of Pascal programs. It is found that some implementations do not permit *integer* or *char* as the base-type but only a subrange of these types.

Set Operators

The relational operators may be used to compare sets. In addition, a number of set operators are available in Pascal. These are listed in Table 6.1. The operators hierarchy is the same as when the symbols are used with arithmetic operations.

Parentheses can be used in set expressions to alter or clarify the evaluation order.

Table 6.1 Pascal Set Operators

Texicographically	Mathematical symbol	Pascal symbol	Type of result
set	{......}	[......]	–
empty set	ϕ	[]	–
union	$\cup$	+	base-type used
difference	–	–	base-type used
intersection	$\cap$	*	base-type used
equality	=	=	boolean
set inequality	$\neq$	< >	boolean
set contains	$\supset$	> =	boolean
set is contained by	$\subset$	< =	boolean
inclusion	$\in$	**in**	boolean

To discuss the operators applicable to set structures the following sets are defined:

 type numbers = **set of** 1 .. 64;
 var a,b,c,d : numbers;
 ....
 a: = [1..64]; b: = [2,8,16,32,64]; c: = [1,5,7,11,13,17];

A set expression can be only assigned to a set variable of the same type.

The relational operators can be used to form a number of boolean expressions using the relational operators. For example:

Expression	Result type	Expression	Result type
a = [1..64]	*true*	c > = [1,5]	*true*
b = a	*false*	a > = b	*true*
b < > c	*true*	b < = c	*false*
b < > [8,16,64,2,32]	*false*	b < = a	*true*

The three operators that specifically operate on sets and produce as a result a new set are:

The union (+) : The union of two sets is a set which contains all the members of both sets
      i.e. d: = b + c + [62,63];
       d is the set [2,8,16,32,64,1,5,7,11,13,17,62,63]

The difference (−) : The difference of two sets is a set which contains all the members of one set that are not members of the other set
      i.e. d: = b − [32,64];
       d is the set [2,8,16]

The intersection (*): The intersection of two sets is a set which contains all the elements that are common to both sets.

> *i.e.* d: = a * b;
> d is the set [2,8,16,32,64]
> d: = b * c;
> d is the empty set [ ]

The inclusion (**in**) : The reserved word **in** can be used to check membership of a particular element in a set
> *i.e.* 64 **in** a   is *true*
> 64 **in** c   is *false*

The reserved word **in** is often used to avoid using cumbersome boolean expressions. It is very useful in constructing boolean functions and for restricting the case selector, thus avoiding errors. For example, the complex boolean expression in Example 4.2 to test for a valid case selector can be simplified with the expression:

> test **in** ['L', 'N', 'E', 'S'];

which is equivalent to:

> (test = 'L') **or** (test = 'N') **or** (test = 'E') **or** (test = 'S')

Other similar examples are:

> (a = x) **and** (b = x) **and** (c = x)   can be rewritten as [a,b,c] = [x]
> (ch > = '0') **and** (ch < = '9')   can be rewritten as ch **in** ['0'..'9']

Worked Example 6.5

Assume a = [1,2,3]   b = [4,5,6]
c = [7,2,3,1]   d = [3,2,1,8]

Evaluate (*i*) (a + b) − c   (*ii*) (a * c) = a
(*iii*) (a + b) * (b − c)   (*iv*) 5 **in** (c − b)
(*v*) c < = d   (*vi*) 8 **in** [1,2,8,8,89]

Solution:

(*i*) [4,5,6]   (*ii*) true
(*iii*) []   (*iv*) false
(*v*) false   (*vi*) true

Worked Example 6.6

Assume that the following two statements are correct Pascal statements, what can you say about these two statements and their data types?

(i) b: = d * a[3];   (ii) b: = d * [3];

Solution:

(i) a is an **array**; b and d could be either *integer* or *real*; b is a variable; d could be either a constant or a variable; the **array** can be of type *real* or *integer*.
(ii) b and d must be sets of type *integer*;

Do not confuse the square brackets used with **array** and **set**.

122

The Record Structure

A few high level languages have the facility of defining data structures consisting of a fixed number of elements which may be of different type. This is in contrast to data structures where all the elements of the structure are of the same type, such as arrays and files. In Pascal the record structure provides this facility. The general form of a record definition is:

    type record-type = **record**
                  field-identifier : data type 1;
                  field-identifier : data type 2;
                      .
                      .
                      .
                  field-identifier : data type n
    end;   Note no matching **begin**.

The description of a record structure, given between **record** and **end**, is called the 'field list' and its components a 'field'. The fields of a record may be of any simple or structured data type, including another **record**. If a **record** includes in its definition another **record** then it is called a 'nested record'. The field-identifiers, for the various fields, must be different from all other field-identifiers within any one record definition.

Simple examples of a record definition:

    type complex = **record**
                    rl : *real*;
                  im : *real*
            end; (∗ complex ∗)

    type WhenTaken = **record**
                  term   : (First, Second, Third);
                  acadyear: (1950 .. 1986)
            end; (∗ WhenTaken ∗)

    type units = **record**
                  unitname: **packed array** [1 .. 20] of *char*;
                  mark   : *integer*;
                  result   : (fail,pass)
            end; (∗ units ∗)

Variable identifiers for record structures are defined as before e.g. for variable identifiers a, b, c of type complex:

    var a, b, c : complex;

Individual elements can be accessed by continuing the variable identifier (as in this example a, b or c) followed by a period (.) and the field identifier (rl or im) i.e.

    a.im   ⎫ are all valid identifiers of type
    b.rl   ⎬ complex which can be assigned
    c.im   ⎭ real values

Worked Example 6.7 Assume that first year undergraduate students enroll in the subjects: Computer Technology (CT105), Engineering (E100), Mathematics (M100) and Physics (P110). Write a Pascal program that accepts numeric examination results for a student and prints his best subject and upon a request for individual results it prints pass for a mark > 50 and fail for less than 50.

```
program exampressure1(input,output);

(* program uses records and sets to allow
   for greater readability                      *)

type subjects = (CT105, E100, M100, P110);
     units = record
                unitname : packed array [1..20] of char;
                mark     : integer;
                result   : (fail, pass);
             end;

var course, bestsubject : subjects;
    unit                : array [subjects] of units;
    subnum              : integer;

begin
  unit[CT105].unitname := 'Computer Tech 105';
  unit[E100].unitname := 'Engineering 100';
  unit[M100].unitname := 'Maths 100';
  unit[P110].unitname := 'Physics 110';

  writeln('Enter marks for CT105, E100, M100 and P110');
  writeln('on separate lines following each prompt');
  writeln;
  for course := CT105 to P110 do        (* enter subject marks and grades *)
  begin
    write('Subject - ', unit[course].unitname,': ');
    readln(unit[course].mark);
    if unit[course].mark < 50 then
      unit[course].result := fail
    else
      unit[course].result := pass
  end; (* for course *)

  bestsubject := CT105;                 (* Determine the best subject *)
  for course := E100 to P110 do
    if (unit[course].mark > unit[bestsubject].mark)
      then bestsubject := course;
  write('Best subject is ');             (* print best subject *)
  writeln(unit[bestsubject].unitname);

  (* request individual results *)
  writeln;
  writeln('Request for individual results');
  writeln('Enter subject number (0 - 3) or exit (4): ');
  read(subnum);
  while ((subnum >= 0) and (subnum <= 3)) do
  begin
    course := CT105;
    while ord(course) < subnum do
      course := succ(course);
    writeln('Grade for ',unit[course].unitname,' is ',unit[course].result);
    writeln('Enter subject number (0 - 3) or exit (4): ');
    read(subnum);
  end; (* while *)
end.

Execution begins...

Enter marks for CT105, E100, M100 and P110
on separate lines following each prompt

Subject - Computer Tech 105    : 12
Subject - Engineering 100      : 34
Subject - Maths 100            : 56
```

```
Subject - Physics 110        : 78
Best subject is Physics 110

Request for individual results
Enter subject number (0 - 3) or exit (4):
0
Grade for Computer Tech 105    is fail
Enter subject number (0 - 3) or exit (4):
1
Grade for Engineering 100      is fail
Enter subject number (0 - 3) or exit (4):
2
Grade for Maths 100            is pass
Enter subject number (0 - 3) or exit (4):
3
Grade for Physics 110          is pass
Enter subject number (0 - 3) or exit (4):
4

Execution terminated.
```

Write a Pascal program that evaluates the input impedance of a ladder network. **Worked Example 6.8**
The components in each branch can be either a resistor, capacitor or an inductor.

```
program RLCladder1(input,output);

(* evaluates a ladder network of RLC components *)

const maxbranches = 20;

type elemtype = record
          component : char;
          value     : real;
               end;
     complex   = record
                   rl,im :real;
                 end;
     complexarray = array [1..20] of complex;

var element : array [1..maxbranches] of elemtype;
    i,branches : integer;
    omega : real;
    z : complexarray;
    zout,otherz : complex;

procedure complexadd(x,y:complex; var z:complex);
begin
  z.rl := x.rl + y.rl;
  z.im := x.im + y.im
end;

procedure complexmult(x,y:complex; var z:complex);
begin
  z.rl := (x.rl * y.rl) - (x.im * y.im);
  z.im := (x.im * y.rl) + (x.rl * y.im)
end;

procedure complexdiv(x,y:complex; var z:complex);
var denominator : real;
```

```pascal
begin
  denominator := sqr(y.rl) + sqr(y.im);
  z.rl := ((x.rl * y.rl) + (x.im * y.im)) / denominator;
  z.im := ((x.im * y.rl) - (x.rl * y.im)) / denominator
end;

function ok(ch : char):boolean;
begin
  ok := ch in ['r','l','c']
end; (* ok *)

procedure enterdata;
begin
  for i := 1 to branches do
    begin
      repeat
        write('Enter name of component ',i,' [r/l/c] = ');
        readln(element[i].component)
      until ok (element[i].component);
      repeat
        write('Enter the value of the component: ');
        readln(element[i].value)
      until element[i].value > 0
    end
end; (* procedure enterdata *)

procedure findz;
var i :integer;
begin
  for i := 1 to branches do
    case element[i].component of
      'r': begin
             z[i].rl := element[i].value;
             z[i].im := 0
           end;
      'c': begin
             z[i].rl := 0;
             z[i].im := -1/(omega * element[i].value)
           end;
      'l': begin
             z[i].rl := 0;
             z[i].im := omega * element[i].value
           end
    end; (* case *)
end;

procedure equivz(branchnum:integer; var outputz:complex);
var tmp1,tmp2,tmp3 : complex;
begin
  if (branchnum = branches) then
    outputz := z[branchnum]
  else
    begin
      equivz(branchnum+2, tmp3);
      complexadd(z[branchnum+1], tmp3, tmp1);
      complexadd(z[branchnum], tmp1, tmp2);
      complexmult(z[branchnum], tmp1, tmp1);
      complexdiv(tmp1, tmp2, tmp1);
      outputz := tmp1;
    end;
end;

begin (* main program *)
  write('Enter number of meshes: ');readln(branches);
  branches := branches * 2;
  enterdata;
  writeln('Enter the frequency at which you wish to evaluate');
  write('the input impedance of the ladder network = ');
  readln(omega);
  findz;
  equivz(2,otherz);
  complexadd(z[1],otherz,zout);
  writeln('Equivalent impedance is ',zout.rl:12:4,' + j ',zout.im:12:4)
end. (* program resistiveladder *)
```

```
Execution begins...

Enter number of meshes: 3
Enter name of component          1 [r/l/c] = r
Enter the value of the component: 1
Enter name of component          2 [r/l/c] = r
Enter the value of the component: 1
Enter name of component          3 [r/l/c] = r
Enter the value of the component: 1
Enter name of component          4 [r/l/c] = r
Enter the value of the component: 1
Enter name of component          5 [r/l/c] = r
Enter the value of the component: 1
Enter name of component          6 [r/l/c] = r
Enter the value of the component: 1
Enter the frequency at which you wish to evaluate
the input impedance of the ladder network = 0
Equivalent impedance is        1.6250 + j        0.0000

Execution terminated.
```

Consider now the following record structure which defines a circuit element connected between two nodes (firstnode and secondnode) within a circuit:

    type CircElem = **record**
                    compontype : **packed array** [1..10] of *char*;
                    frequency : array [1..100] of *real*;
                    compvalue : *real*;
                    firstnode, secondnode : *integer*
              end; (∗ CircElem ∗)

Assume that a variable element1 has been declared as being of type CircElem, then the assignments

    element1.compontype := 'Capacitor ';
    element1.compvalue := 10E-6
    element1.frequency[5] := 1000.0;
    element1.firstnode := 2;
    element1.secondnode := 7;

describe a capacitor of value $10\mu F$ at 1000Hz which is connected between nodes 2 and 7 in the circuit.

If two record variables have identical record definitions, then it is possible to assign all the fields of one record variable to another using only one assignment statement:

    element2: = element1;

Pascal also provides the reserved word **with** which reduces the tedium of accessing individual components of records. Record variables can be accessed as simple variables using **with**. This method reduces the length of a program and improves readability:

    with element1 **do**
      begin
          comportype := 'Resistor'
          compvalue := 2000.0
      for i = 1 **to** 100 **do**

```
        frequency[i] : 0.0;
        firstnode    := 3;
        secondnode   := 8
    end;
```

A comparison between the **array** and **record** data structures is appropriate here:

array	**record**
(i) fixed length	(i) fixed length
(ii) elements of the same data type	(ii) components of different data type
(iii) nesting permitted	(iii) nesting permitted
(iv) random access possible	(iv) random access possible
(v) individual element can be selected with the aid of a subscript	(v) individual components can be selected by individually naming them

In some cases records which vary slightly in their definition are required. Pascal allows this by using variant record definition.

Variant Record

The record defined above has a fixed number and type of fields. Pascal allows a record to have a 'variant part', i.e. the effective number and type of fields may change during the running of the programme. A variant record definition has a fixed-part (if there is one) followed by a variant-part. The declaration of a field whose values determine which particular variant of the record is used is called the 'tag field'.

Consider the example given previously for units:

```
type units = record
        unitname : packed array [1..20] of char;
        mark     : integer;
        result   : (fail,pass);
    end;
```

and assume that now we required a more descriptive answer for the result rather than just fail or pass.

We can write a variant record as

```
type failpass = (fail,pass);
     passes   = (distinction,credit,straightpass);
     fails    = (repeat,expel);
     units    = record
            unitname : packed array[1..20] of char;
            mark     : integer;
            case result : failpass of
            pass : (passmark : passes);
            fail : (failmark : fails);
        end;
```

In this example, the tag field is failpass which must be of ordinal type and, each value of its components (pass or fail) must appear exactly once as a case label in the variant part.

Repeat example 6.8 using variant records. **Worked Example 6.9**

```
program exampressure3(input,output);

(* This version of exampressure uses variant records
   as form of data storage                           *)

type subjects = (CT105, E100, M100, P110);
     failpass = (fail, pass);
     passes   = (distinction, credit, straightpass);
     fails    = (repeatagain, expel);
        units = record
                  unitname : packed array [1..20] of char;
                  mark     : integer;
                  case result : failpass of
                       pass : (passmark : passes);
                       fail : (failmark : fails);
                end;

var course, bestsubject : subjects;
    unit                : array [subjects] of units;
    subnum              : integer;
begin
  unit[CT105].unitname := 'Computer Tech 105';
  unit[E100].unitname := 'Engineering 100';
  unit[M100].unitname := 'Maths 100';
  unit[P110].unitname := 'Physics 110';

  writeln('Enter marks for CT105, E100, M100 and P110');
  writeln('on separate lines following each prompt');
  writeln;
  for course := CT105 to P110 do      (* enter subject marks and grades *)
  begin
    write('Subject - ', unit[course].unitname,': ');
    readln(unit[course].mark);
    if unit[course].mark < 50 then
       unit[course].result := fail
    else unit[course].result := pass;
    if unit[course].mark < 30 then
       unit[course].failmark := expel
    else if unit[course].mark < 50 then
       unit[course].failmark := repeatagain
    else if unit[course].mark < 65 then
       unit[course].passmark := straightpass
    else if unit[course].mark < 75 then
       unit[course].passmark := credit
    else if unit[course].mark < 100 then
       unit[course].passmark := distinction;
  end; (* for course *)

  bestsubject := CT105;                 (* Determine the best subject *)
  for course := E100 to P110 do
    if (unit[course].mark > unit[bestsubject].mark)
       then bestsubject := course;
  write('Best subject is ');            (* print best subject *)
  writeln(unit[bestsubject].unitname);

(* request individual results *)
  writeln;
  writeln('Request for individual results');
  writeln('Enter subject number (0 - 3) or exit (4): ');
  read(subnum);
  while ((subnum >= 0) and (subnum <= 3)) do
  begin
    course := CT105;
    while ord(course) < subnum do
      course := succ(course);
```

129

```
            write('Grade for ',unit[course].unitname,' is ');
            if unit[course].result = fail then
              writeln(unit[course].failmark)
            else
              writeln(unit[course].passmark);
            writeln('Enter subject number (0 - 3) or exit (4): ');
            read(subnum);
     end; (* while *)
end.

Execution begins...

Enter marks for CT105, E100, M100 and P110
on separate lines following each prompt

Subject - Computer Tech 105    : 12
Subject - Engineering 100      : 34
Subject - Maths 100            : 56
Subject - Physics 110          : 78
Best subject is Physics 110

Request for individual results
Enter subject number (0 - 3) or exit (4):
0
Grade for Computer Tech 105    is expel
Enter subject number (0 - 3) or exit (4):
1
Grade for Engineering 100      is repeatagain
Enter subject number (0 - 3) or exit (4):
2
Grade for Maths 100            is straightpass
Enter subject number (0 - 3) or exit (4):
3
Grade for Physics 110          is distinction
Enter subject number (0 - 3) or exit (4):
4

Execution terminated.
```

Worked Example 6.10 Using records define complex numbers and perform complex addition, complex multiplication and complex division for any two given complex numbers.

Solution:

```
program ComplexRecords(input,output);

(* This program creates procedures for the
   implementation of complex arithmetic      *)

type complex = record
                  rl : real;
                  im : real;
               end;
var a,b,c : complex;

procedure ComplexAdd(x,y:complex; var z:complex);
begin
  z.rl := x.rl+y.rl;
  z.im := x.im+y.im
end (* ComplexAdd *);

procedure ComplexMult(x,y:complex; var z:complex);
begin
  z.rl := (x.rl*y.rl)-(x.im*y.im);
  z.rl := (x.im*y.rl)+(x.rl*y.im)
end (* ComplexMult *);

procedure ComplexDiv(x,y:complex; var z:complex);
var denominator : real;
```

```
begin
  denominator := sqr(y.rl)+sqr(y.im);
  z.rl := ((x.rl*y.rl)+(x.im*y.im))/denominator;
  z.im := ((y.rl*x.im)-(x.rl*y.im))/denominator
end (* ComplexDiv *);

begin
  write('Enter complex real and imaginary parts =');
  readln(a.rl,a.im);
  write('Enter complex real and imaginary parts =');
  readln(b.rl,b.im);
  ComplexAdd(a,b,c);
  writeln('Complex addition : real = ',c.rl:5:2,' imag = ',c.im:5:2);
  ComplexMult(a,b,c);
  writeln('Complex multiplication : real = ',c.rl:5:2,' imag = ',c.im:5:2);
  ComplexDiv(a,b,c);
  writeln('Complex division : real = ',c.rl:5:2,' imag = ',c.im:5:2);
end.

Execution begins...

Enter complex real and imaginary parts = -1 1
Enter complex real and imaginary parts = 1 2
Complex addition : real =  0.00 imag =  3.00
Complex multiplication : real = -1.00 imag =  3.00
Complex division : real =  0.20 imag =  0.60

Execution terminated.
```

The Pointer Data Type

So far variables have been encountered which are declared in the **var** declaration part of the program which determines their type and their identifier. The **var** declaration also creates these variables *i.e.* allocates memory space (called the stack) which remains in existence during the execution of the block in which the variables are declared. For example, when a variable XX of type *real* is declared, a store location named XX is allocated which contains the *real* data as shown in Fig. 6.1.

Pointers are considered beyond the scope of this introductory book. For more details on pointers see the first four books in References.

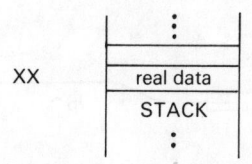

```
          ┌───────────┐
          │    ⋮      │
    XX    │ real data │
          │   STACK   │
          │    ⋮      │
          └───────────┘
```

Fig. 6.1 Memory allocation of real variable XX.

These variables cease to exist when the computation exits from the block in which they have scope. We refer to such variables as 'static' variables.

However, it is not always possible to determine in advance the storage requirements of a data structure. A data structure which expands and contracts or can be even disposed of during program execution may be required. Pascal allows such 'dynamic' variables to be created.

As usual any new type introduced into a Pascal program is defined in the **type** definition part. A pointer-type data structure can be defined as:

type pointer-type = ^base-type

The pointer-type is a dynamic data type whose dynamic variables (pointers) 'reference' or 'point to' objects of the specified type-identifier. The base-type can be any Pascal simple or structured data type. An up-arrow (↑) or circumflex (^) precedes the base-type and it is this symbol which informs us and the compiler that we are dealing with pointer data types. The above definition also indicates that the pointer-type is 'associated' or 'bound' to the base-type and this fact is used by the compiler for type checking. A typical example of this type of definition is:

type realpointer = ^*real*;

Any variable of pointer-type realpointer can be declared as:

var rp : realpointer;

Normally, variables created using the above declaration are defined as static variables (see XX above). But the above declaration involves a dynamic variable and the **var** declaration has the following interpretation: The dynamic variable (rp) is associated to the pointer-type realpointer and is also given the capability of creating variables *i.e.* to be allocated memory space. rp^ is the actual *real* value being pointed to. rp^ is treated as any other variable *e.g.*

rp^ := 25.5; (∗ assigns the value 25.5 to the variable pointed to by the value of rp ∗)

The capability of creating variables is realised by calling the standard procedure *new*. The call

new(rp);

has the effect of allocating space and assigns its address to rp. The memory space allocated to dynamic variables is called the 'heap'. The value of the data pointed to can be accessed by following the pointer variable rp with a circumflex (rp^) *i.e.* rp^ is the 'address' of the heap location which contains the *real* value. Fig. 6.2 illustrates the relationship between rp and rp^.

address is the computer's internal notation for a particular location in memory.

The physical significance of computer addresses is discussed in Downton, A. C. *Computers and Microprocessors: components and systems* (Van Nostrand Reinhold (UK), 1984), Chapter 2.

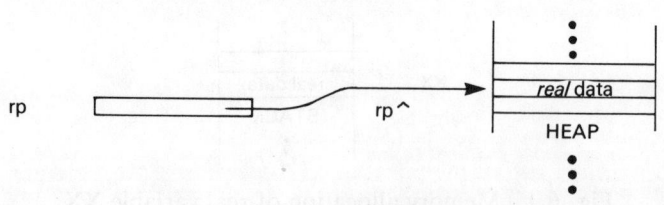

Fig. 6.2 Relationship between a pointer and its data.

The heap memory space acquired can be disposed of by calling another standard Pascal procedure *dispose*:

dispose(rp); (∗ releases previously acquired memory space ∗)

To maintain storage economy the programmer must ensure that all unwanted pointer variables are disposed.
dispose(rp) may dispose all storage allocations on the heap even after rp!

The value of a pointer (rp) cannot be printed or inspected. It can only be compared for equality (=) and inequality (<>) to the value of another pointer variable. Pointer variables may be also assigned, and passed as subprogram parameters. Every pointer-type includes a special value that is denoted by the reserved word **nil**. This is used to indicate that a variable does not reference a location but it exists. It is useful for checking the existence of heap allocation for a variable *i.e.*

    if rp   = **nil**
       then *writeln*('No pointer allocation')
       else *writeln*('Heap has been allocated')

Pointer variables are usually used with records and very seldom are of simple type such as *real*, *integer* or *char*. An example using pointers with a simple type *char* is shown here:

```
program pointertest(input,output);

(* tests the use of pointers *)

type string = array [1..40] of char;

var ptr1,ptr2: ^string;

begin
  new(ptr1);
  new(ptr2);
  ptr1^ := 'This is a string pointed to by ptr1';
  ptr2 := ptr1;
  writeln('ptr1 and ptr2 now point to this string:');
  writeln(ptr1^);writeln;
  writeln('We now change ptr1 ...');
  ptr1^ := 'This is a new string for  our  ptr1';
  writeln('This is what ptr2 points to: ');
  writeln(ptr2^);
end.

Execution begins...

ptr1 and ptr2 now point to this string:
This is a string pointed to by ptr1

We now change ptr1 ...
This is what ptr2 points to:
This is a new string for  our  ptr1

Execution terminated.
```

A few key points that you should note when dealing with pointers are:

(i) Two pointer expressions can be tested for equality.
(ii) Pointer variables allow memory stored to be shared, i.e. several pointers can point to the same object.
(iii) A pointer variable cannot point to any named variable declared in the program.
(iv) Dynamic variables are created and destroyed dynamically during the execution of the program.
(v) Dynamic variables are not referred to by user-defined identifiers; they are referenced by the use of pointer variables which point to the dynamically created variables.

Repeat example 6.8 using pointers.

Worked Example 6.12

```
program RLCladder(input,output);

(* program simulates a ladder network of RLC components *)

const maxbranches = 20;

type elemtype = record
        component : char;
        value     : real;
              end;
     complex  = record
                  rl,im :real;
                end;
     complexarray = array [1..20] of complex;

var element : array [1..maxbranches] of ^elemtype;
    i,branches : integer;
    omega : real;
    z : complexarray;
    zout,otherz : complex;

procedure complexadd(x,y:complex; var z:complex);
begin
  z.rl := x.rl + y.rl;
  z.im := x.im + y.im
end;

procedure complexmult(x,y:complex; var z:complex);
begin
  z.rl := (x.rl * y.rl) - (x.im * y.im);
  z.im := (x.im * y.rl) + (x.rl * y.im)
end;

procedure complexdiv(x,y:complex; var z:complex);
var denominator : real;
begin
  denominator := sqr(y.rl) + sqr(y.im);
  z.rl := ((x.rl * y.rl) + (x.im * y.im)) / denominator;
  z.im := ((x.im * y.rl) - (x.rl * y.im)) / denominator
end;

function ok(ch : char):boolean;
begin
  ok := ch in ['r','l','c']
end; (* ok *)

procedure enterdata;
begin
  for i := 1 to branches do
   begin
    repeat
      write('Enter name of component ',i,' [r/l/c] = ');
      readln(element[i]^.component)
    until ok (element[i]^.component);
    repeat
      write('Enter the value of the component: ');
      readln(element[i]^.value)
    until element[i]^.value > 0
   end
end; (* procedure enterdata *)

procedure findz;
var i :integer;
begin
  for i := 1 to branches do
    case element[i]^.component of
      'r': begin
             z[i].rl := element[i]^.value;
             z[i].im := 0
           end;
      'c': begin
             z[i].rl := 0;
             z[i].im := -1/(omega * element[i]^.value)
           end;
```

```
      'l': begin
             z[i].rl := 0;
             z[i].im := omega * element[i]^.value
           end
    end; (* case *)
end;

procedure equivz(branchnum:integer; var outputz:complex);
var tmp1,tmp2,tmp3 : complex;
begin
  if (branchnum = branches) then
    outputz := z[branchnum]
  else
    begin
      equivz(branchnum+2, tmp3);
      complexadd(z[branchnum+1], tmp3, tmp1);
      complexadd(z[branchnum], tmp1, tmp2);
      complexmult(z[branchnum], tmp1, tmp1);
      complexdiv(tmp1, tmp2, tmp1);
      outputz := tmp1;
    end;
end;

begin (* main program *)
  write('Enter number of meshes: ');readln(branches);
  branches := branches * 2;
  for i := 1 to branches do
    new(element[i]);
  enterdata;
  writeln('Enter the frequency at which you wish to evaluate');
  write('the input impedance of the ladder network = ');
  readln(omega);
  findz;
  equivz(2,otherz);
  complexadd(z[1],otherz,zout);
  writeln('Equivalent impedance is ',zout.rl:12:4,' + j ',zout.im:12:4)
end. (* program resistiveladder *)

Execution begins...
Enter number of meshes: 3
Enter name of component         1 [r/l/c] = r
Enter the value of the component: 1
Enter name of component         2 [r/l/c] = r
Enter the value of the component: 1
Enter name of component         3 [r/l/c] = r
Enter the value of the component: 1
Enter name of component         4 [r/l/c] = r
Enter the value of the component: 1
Enter name of component         5 [r/l/c] = r
Enter the value of the component: 1
Enter name of component         6 [r/l/c] = r
Enter the value of the component: 1
Enter the frequency at which you wish to evaluate
the input impedance of the ladder network = 0
Equivalent impedance is       1.6250 + j       0.0000

Execution terminated.
```

Summary

Pascal has four structured data types: **array, file, set** and **record**. Each structure is made up of one or more component values of simple data type. It is possible to form even more complex structures by using structured data types as their constituent components. Each data structure has a specified method for accessing and storing its component values. A number of standard Pascal subprograms are available for use with structured data types.

The array is the simplest and most widely used data structure. Array variables can store a fixed number of data values all of which must be of the same type. The data values stored can be of any simple or structured data type. To access a particular element, specify the array name followed by a subscript inside square brackets.

Array values can be stored in memory in two forms: unpacked and packed. A sequence of **packed array** elements of *char* type is defined as a string. Strings are useful in handling character variables as a single entity.

Only sequential files are available with Pascal which can either be internal or external. All file components within a file must be of the same type but, unlike an array, the file size can be *unlimited* (typical limit is 2^{10}). Files with different types of components can be created; files with *char* values are predefined in Pascal as *text*. When a file is created a file-buffer is automatically available which allows us to see and use one file component at a time.

Pascal allows operations with sets which are used whenever we are not concerned with the order in which elements are stored or the number of times elements are repeated in a set. Values in a set must be ordinal type within an implementation-defined limit (typical limit is 2^7). Individual elements of a set can be accessed directly by name without the need of a subscript or a file-buffer. Sets are extremely useful in forming boolean expressions.

Records are used whenever a structure which consists of a fixed number of elements is required but not necessary all having the same data type. Each type definition within a record is called a field. A field-identifier can be accessed using the period-notation or the reserved word **with**. Records are extensively used in commercial programming.

Pointer variables contain the address of memory locations where data values are stored. Pointers are very useful when large amounts of data are stored in memory and for a certain reason we want to access some of the data in a non sequential order. Instead of moving the stored data values, pointers are used to point at the required memory locations.

Problems

6.1 Assume a = [4,5,57]   b = [6,5,4]
         c = [57,3,2,1]   d = [3,3,4,9]

   Evaluate (i)   (a+b)−c   (ii) (a∗c)=b
             (iii) (a+c)−(b∗c)   (iv) (a+b+c)∗d
             (v)   (b+d)−c   (vi) (a+b)+(b+c)

6.2 Using the definition

   type complex = record
                    rl:real;
                    im:real
                  end;
   Write procedures to (i)   subtract two complex numbers
                       (ii) to calculate the square root of a complex number
                       (iii) to raise a complex number to a given integer power.

6.3 Suppose that a file called circuit is declared as:

```
    type  component = record
                        name:packed array [1..10] of char;
                        value:real;
                        n1,n2:integer
                      end;
    var  circuit: file of component;
```
Write a program to print a listing of the file circuit.

6.4 Write a sorting procedure using any sorting method except Quick Sort.

6.5 Write an editor program to edit a file with the following commands:

   'D' = to delete a line
   'I' = to insert a line
   'M' = to modify a character
   'S' = to substitute a string with another string.

6.6 Write a program to read an unsorted sequence of integer numbers in the range −45 to +45. As the numbers are read the program should sort them using two stacks: stackpositive and stacknegative. After processing the last number, print the sorted series.

6.7 A linear T-section circuit shown in Fig. 6.3 can be converted to an equivalent π-section using the following expressions.

$$Y_A = G_A + jB_A = \frac{Z_2}{Z}$$

$$Y_B = G_B + jB_B = \frac{Z_3}{Z}$$

$$Y_C = G_C + jB_C = \frac{Z_1}{Z}$$

where $Z = Z_1 Z_2 + Z_1 Z_3 + Z_2 Z_3$
and $Z_1 = R_1 + jX_1$

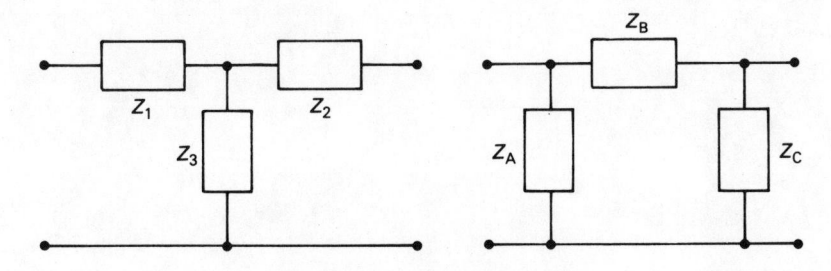

Fig. 6.3 T and π-sections.

$Z_2 = R_2 + jX_2$
$Z_3 = R_3 + jX_3$

Write a Pascal program to perform this conversion

6.8 Write a Pascal program to convert a given π-section into a T-section.

6.9 In a single program incorporate the π to T transformation and the T to π transformation giving the user the option of transformation.

6.10 Write a Pascal program to calculate the input resistance of a resistive ladder network. Test your problem using the following circuit.

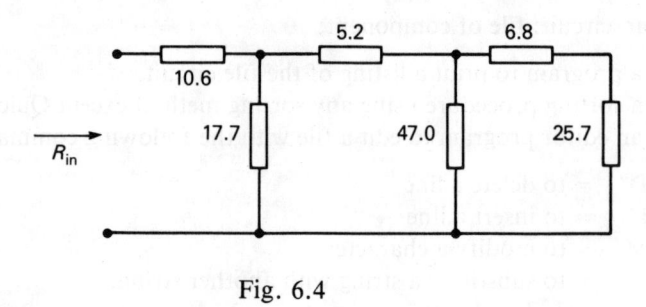

Fig. 6.4

6.11 Repeat Problem 6.10 using impedances instead of resistances.

Case Studies 7

□ To write a program to calculate the magnitude and phase characteristics of any transfer function given its pole-zero locations. □ To write a program to calculate the element values of a band pass active filter. □ To write a program for the analysis of passive *RLC* circuits.	**Objectives**

Network Transfer Functions

Network functions can be expressed as the ratio of two polynomials in the complex frequency variable s:

Kuo, F.F. Network *Analysis and Synthesis* (Wiley, 1966).

$$H(s) = \frac{N(s)}{D(s)} = \frac{a_m s^m + a_{m-1} s^{m-1} + \dots + a_0}{b_n s^n + b_{n-1} s^{n-1} + \dots + b_0} \tag{7.1}$$

where H(s) is a rational function of s with m and n having integer values (n > m). The coefficients of both numerator N(s) and denominator D(s) polynomials are real for networks composed of passive elements and with no controlled sources. Equation 7.1 can be rewritten in factored form as:

$$H(s) = \frac{N(s)}{D(s)} = \frac{K(s-z_1) \cdot (z-z_2) \dots (s-z_m)}{(s-p_1) \cdot (s-p_2) \dots (s-p_n)} = K \frac{\prod_{i=1}^{m}(s-z_i)}{\prod_{j=1}^{n}(s-p_j)} \tag{7.2}$$

where K is a scale factor and z_i and p_j are the roots of the numerator and denominator polynomials respectively. Since the coefficients of H(s) are all real then z_i and p_j must occur in real, imaginary and/or complex conjugate pair form. When the complex frequency variable s takes on one of the values $z_1 \dots z_m$ the network function described by Equation 7.2 becomes zero; such complex frequencies are called 'zeros'. When s is assigned values $p_1 \dots p_n$ then the network function becomes infinite; such frequencies are called 'poles'.

Given a network transfer function in terms of its poles and zeros it is possible to calculate the network amplitude and phase response using a graphical method. Assume the given transfer function is:

$$H(s) = \frac{K(s-z_1) \cdot (s-z_2)}{(s-p_1) \cdot (s-p_2) \cdot (s-p_3)} \tag{7.3}$$

In practice, we are interested in the steady state response of networks to an applied sinusoid, therefore, we let the complex frequency variable be $s = j\omega$ i.e. equation 7.3 can be expressed as

$$(Hj\omega) = \frac{K(j\omega - z_1) \cdot (j\omega - z_2)}{(j\omega - p_1) \cdot (j\omega - p_2) \cdot (j\omega - p_3)} \tag{7.4}$$

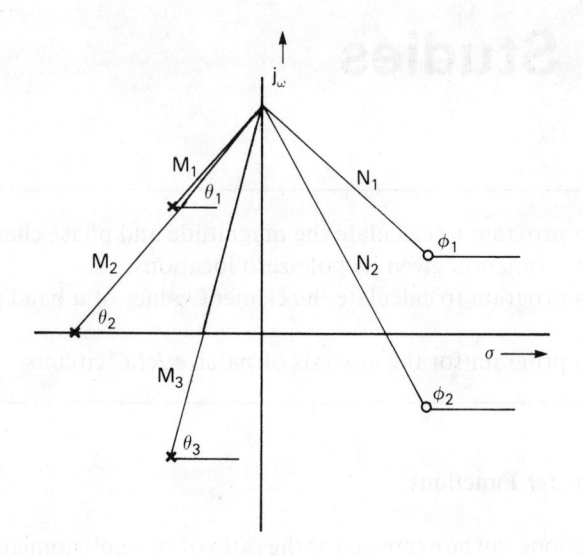

Fig. 7.1

Each factor of Equation 7.4 corresponds in the complex plane to a vector from z_i or p_j to any point $j\omega$ on the imaginary axis as shown in Fig. 7.1. Using the notation from Fig. 7.1 the transfer function can be written as

$$H(j\omega) = \frac{KN_1N_2}{M_1M_2M_3} e^{j(\phi_1 + \phi_2 - \theta_1 - \theta_2 - \theta_3)} \quad (7.5)$$

Thus the magnitude response $M(\omega)$ can be expressed as

$$M(\omega) = \frac{K\,N_1N_2}{M_1M_2M_3} \quad (7.6)$$

and the phase response as

$$\phi = \phi_1 + \phi_2 - \theta_1 - \theta_2 - \theta_3 \quad (7.6)$$

Transfer Function Analysis Program

```
program Transfer(input, output);

type complex = record
                 rl :real;
                 im :real
               end;
     complexarray = record
                      rl :array [1..20] of real;
                      im :array [1..20] of real
                    end;
var npoints, Npoles, Nzeros :integer;
    LinLog, RatdB, DegRad, YN :char;
    t1, t2, dummy, logstep, linstep :real;
    poles, zeros :complexarray;

procedure space( x: integer);
var i: integer;
begin
  for i := 1 to x do
    write(' ');
end; (space)
```

```
procedure nl(x:integer);
var i : integer;
begin
for i := 1 to x do writeln;
end; {nl}

function log(x:real):real;
begin
log := ln(x)/ln(10);
end; {log}

function modulus(position :complex; freq:real):real;
begin
  modulus := sqrt(sqr(freq-position.im)+sqr(position.rl))
end; { modulus}

function phase(position :complex; freq :real; DegRad :char):real;
const pi = 3.1414592653589;
var phi,efap  :real;
begin
  position.im := freq-position.im;
  position.rl := -position.rl;
  if position.rl = 0.0 then if position.im > 0.0 then efap := pi/2
                                                else efap := -pi/2;
  if position.im = 0.0 then if position.rl >=0.0 then efap := 0.0
                                                else efap := pi;
  if ((position.rl <> 0.0) and (position.im <> 0.0))
    then begin
      phi := arctan(abs(position.im/position.rl));
      if position.rl > 0.0 then if position.im >0.0 then efap := phi
                                                    else efap := -phi
                           else if position.im >0.0 then efap := pi-phi
                                                    else efap :=-pi+phi
    end;
  if DegRad = 'D' then phase := efap * 180 /pi
             else phase := efap;
end; {phase}

procedure pzinput(number :integer; var pz:complexarray);
var i: integer;
begin
  i := 1;
  while not (i > number) do
    begin
      write ('Enter real and imag part = ');
      readln(pz.rl[i],pz.im[i]);
      if pz.im[i] <> 0.0 then
      begin
        pz.rl[i+1] := pz.rl[i];
        pz.im[i+1] := -pz.im[i];
        i := i +1;
      end;
      i := i + 1;
    end;
end; {pzinput}

procedure pzoutput(number: integer; pz:complexarray; ch:char);
var i :integer;
begin
  nl(2);
  space(16);write('The ');
  if ch = 'P' then writeln('poles are ')
              else writeln('zeros are ');
  space(10);write('Real Part');space(5);write('Imag Part');
  nl(2);
  for i := 1 to number do
  begin
    space(9);write(pz.rl[i]:6:3);space(8);
    writeln(pz.im[i]:6:3);
  end;
  nl(2);
end; {pzoutput}

procedure freqspecs(var f1,f2:real; var LinLog, RatdB, DegRad:char);
```

There is only a need to enter one of the complex conjugate pair of a pole or a zero. The procedure pzinput generates the other one.

```
begin
  nl(2);writeln(' ** Frequency Specifications ** ');
  nl(2);
  repeat
    write('Enter upper and lower frequency limits in rad/s = ');
    readln(f1,f2);
  until ((f1 >= 0.0) and (f2>0.0));
  if f1>f2 then begin
    dummy := f2;
    f2 := f1;
    f1 := dummy;
  end;
  repeat
    write('Frequency response in linear or log form ( enter I or G ) = ');
    readln(LinLog);
  until LinLog in ['I', 'G'];
  if LinLog = 'I' then begin
    write('Enter frequency step in rad/s = ');
    readln(linstep);
    if f1 = 0.0 then f1 := 1E-9;
    npoints := round((f2-f1)/linstep)+1;
  end
       else begin
    write('Enter number of points for log scale = ');
    readln(npoints);
    if f1 = 0.0 then if f2 > 1.000001 then f1 := 1/f2
                                     else f1 := 1E-9;
    dummy := (ln(f2/f1)/ln(10))/(npoints-1);
    logstep := exp(dummy*ln(10))
  end;
  repeat
    write('Output in d(Bs or in R(atio = ');
    readln(RatdB);
  until RatdB in ['d', 'R'];
  repeat
    write('Phase In D(egress or R(adians = ');
    readln(DegRad);
  until DegRad in ['D','R']
end; {freqspecs}

function AmpRes(number:integer; pz:complexarray; freq:real):real;
var i :integer;
    polzer :complex;
    Amplit :real;
begin
  if number = 0 then AmpRes := 1.0
       else begin
         Amplit := 1.0;
         for i := 1 to number do
         begin
           polzer.rl := pz.rl[i];
           polzer.im := pz.im[i];
           Amplit := Amplit*modulus(polzer,freq);
         end;
         AmpRes := Amplit;
       end;
end; {ampres}

function PhaRes(number:integer; pz:complexarray; freq:real;
                DegRad:char) :real;
var i :integer;
    psi :real;
    polzer: complex;
begin
  if number = 0 then PhaRes := 0.0
       else begin
         psi := 0.0;
         for i := 1 to number do
         begin
           polzer.rl := pz.rl[i];
           polzer.im := pz.im[i];
           psi := psi + phase(polzer,freq,DegRad)
         end;
         PhaRes := psi;
       end;
end; {phares}
```

Just in case the frequency values are entered in the wrong order.

```
procedure TableFreq;
var i: integer;
  omega,theta,mag,dc,PhaZer,PhaPol,DisPol,DisZer :real;
begin
  space(20);
  writeln('Frequency Response');
  nl(1);
  space(5);write('Frequency');space(13);write('Magnitude');
  space(14);writeln('Phase');
  space(6);write('rad/s');space(18);
  if RatdB = 'd' then write('dB')
                 else write ('Ratio');
  space(16);
  if DegRad = 'D' then write('degrees')
                  else write('radians');
  nl(1);
  dc:= 1/(AmpRes(Nzeros,zeros,0.0)/AmpRes(Npoles,poles,0.0));
  omega := f1;
  for i := 1 to npoints do
  begin
    DisPol := AmpRes(Npoles,poles,omega);
    DisZer := AmpRes(Nzeros,zeros,omega);
    if DegRad = 'D' then begin
      PhaPol := PhaRes(Npoles, poles, omega, 'D');
      PhaZer := PhaRes(Nzeros, zeros, omega, 'D');
    end
    else begin
      PhaPol := PhaRes(Npoles, poles, omega ,'R');
      PhaZer := PhaRes(Nzeros, zeros, omega ,'R');
    end;
    if RatdB = 'R' then mag := dc*DisZer/DisPol
                   else mag := 20*log(dc*DisZer/DisPol);
    theta := PhaZer-PhaPol;
    write(omega:10:3,' ':13,mag:10:3);
    writeln(' ':13,theta:10:3);
    if LinLog = 'l' then omega := omega +linstep
                    else omega := omega +logstep
  end
end; {tabfreq}

{main program starts here}
begin
  repeat
    write('Enter number of zeros = ');readln(Nzeros);nl(1);
    if Nzeros <> 0 then pzinput(Nzeros, zeros);nl(1);
    write('Enter number of poles = ');readln(Npoles);
    pzinput(Npoles,poles);
    nl(1);
    if Nzeros <> 0 then pzoutput(Nzeros, zeros, 'Z')
                   else writeln('No zeros');
    nl(1);pzoutput(Npoles, poles, 'P');
    repeat
      freqspecs(f1,f2,LinLog, RatdB, DegRad);
      TableFreq;
      write('? rerun with different frequency specs y/n = ');
      readln(YN);
    until YN in ['N', 'n'];
    write('? rerun program y/n = '); readln(YN);
  until YN in ['N', 'n'];
end.
```

Calculation of magnitude and phase.

Execution begins...

Enter number of zeros = 0

Enter number of poles = 3
Enter real and imag part = -0.5 -0.866
Enter real and imag part = -1.0 0

No zeros

            The poles are
        Real Part   Imag Part

143

```
                    -0.500            -0.866
                    -0.500             0.866
                    -1.000             0.000


       ** Frequency Specifications **

       Enter upper and lower frequency limits in rad/s = 0.0 10.0
       Frequency response in linear or log form ( enter I or G ) = I
       Enter frequency step in rad/s = 1.0
       Output in d(Bs or in R(atio = d
       Phase In D(egress or R(adians = D
                         Frequency Response

             Frequency             Magnitude              Phase
              rad/s                   dB                 degrees
              0.000                  0.000               -0.000
              1.000                 -3.011             -135.008
              2.000                -18.130             -209.754
              3.000                -28.634             -231.019
              4.000                -36.125             -241.043
              5.000                -41.939             -246.932
              6.000                -46.690             -250.821
              7.000                -50.706             -253.584
              8.000                -54.186             -255.649
              9.000                -57.255             -257.252
             10.000                -60.000             -258.532
       ? rerun with different frequency specs y/n = n
       ? rerun program y/n = n

       Execution terminated.
```

Attikiouzel, J. and Linggard, R. Low Cost Active Low Pass and Band Pass Network, *International Journal of Circuit Theory and Applications,* Vol. 2, No. 4, pp. 397–400, 1974. An introductory treatment of operational amplifiers is given in Horrocks, D. H. *Feedback Circuits and Op. Amps* (Van Nostrand Reinhold (UK), 1983).

Active Filter Synthesis

Active RC filters are composed of resistors, capacitors and active devices. In practice, the active device is an operational amplifier. By appropriate choice of network configuration low-pass, high-pass or band-pass responses can be obtained. For example, using the circuit configuration shown in Fig. 7.2 a second order low-pass filter response can be obtained as

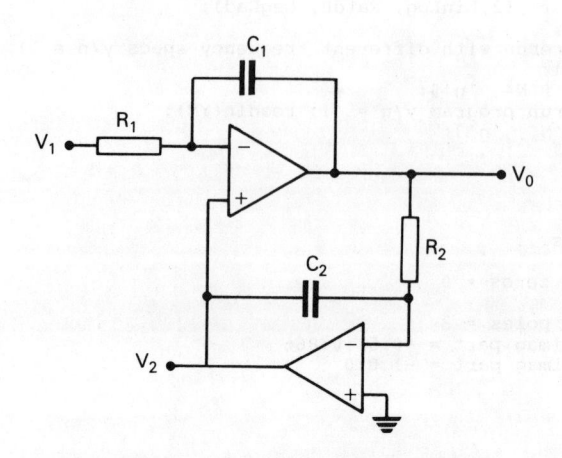

Fig. 7.2 Second Order building block.

$$\frac{V_2}{V_1} = \frac{\frac{1}{C_1 C_2 R_1 R_2}}{s^2 + s\frac{1}{R_2 C_2} + \frac{1}{C_1 C_2 R_1 R_2}} \tag{7.8}$$

Higher order filter responses of even order can be obtained by cascading a number of these functions; odd order filters can be obtained by cascading a first order *RC*-section with the required number of second order sections.

As discussed in the first section of this chapter a network function can be written in terms of its poles and zeros. If all-pole network functions are considered now then a network function given in Equation 7.1 can be decomposed into a number of second order functions as

$$H(s) = K \cdot \frac{1}{s^2 + \alpha_1 s + \beta_1} \cdot \frac{1}{s^2 + \alpha_2 s + \beta_2} \cdots \tag{7.9}$$

i.e. each pair of complex poles may be realised by one active *RC* second order stage. The element values can be obtained by equating coefficients in Equations 7.8 and 7.9.

Using the circuit given in Fig. 7.2 it is also possible to obtain a band-pass response if we consider

$$\frac{V_0}{V_1} = \frac{\frac{-s}{R_1 C_1}}{s^2 + s\frac{1}{R_2 C_2} + \frac{1}{C_1 C_2 R_1 R_2}} \tag{7.10}$$

$$= \frac{-Ks}{s^2 + \frac{1}{Q}s + \omega_0^2}$$

where Q is the Q-factor and ω_0 is the resonant frequency. For this circuit

$$\omega_0 = \frac{1}{\sqrt{(R_1 R_2 C_1 C_2)}}$$

$$Q = \sqrt{\left(\frac{R_2 C_2}{R_1 C_1}\right)} \tag{7.11}$$

Filter specifications (pole-zero) are usually given in terms of a low-pass normalised filter. The band-pass pole-zero locations can be obtained using the transformation

$$s = \frac{\omega_0}{\text{BW}}\left[\frac{s}{\omega_0} + \frac{\omega_0}{s}\right] \tag{7.12}$$

where ω_0 is the resonant frequency and BW is the bandwidth, *i.e.* for each pair of complex conjugate low-pass pole locations we obtain four band-pass poles and two zeros at the origin. If there is a real low-pass pole then using Equation 7.12 a band-pass complex conjugate pair is obtained.

Active Circuit Synthesis Program

In this program band-pass filters are realised given the upper and lower cut-off frequencies. The band-pass filter can be specified to be of Butterworth or Chebyshev type. The program first calculates the low-pass pole locations from which the band-pass pole-zero locations are calculated using the transformation given in Equation 7.12. Second order sections are then formed and the coefficients are equated with Equation 7.10 in turn to obtain the element values.

```
program ActiveFilterDesign(input,output);

const pi = 3.141592653589;

type Xarray = array [1..20 ] of real;

var n :integer;
    BW, WC :real;
    relp, imlp, rebp1, imbp1, rebp2, imbp2 :Xarray;
    bp1, bp2, cp1, cp2 :Xarray;

procedure space(x:integer);
var i :integer;
begin
    for i := 1 to x do
        write(' ');
end ( space );

procedure nl(x:integer);
var i :integer;
begin
    for i := 1 to x do
        writeln
end ( nl );

procedure Butt(n:integer; var sigma,omega:Xarray);
var l,k :integer;
    a :real;
begin
    for k := 1 to 2 * n do
    begin
        a := ((2 * k - 1)/(2 * n)) * pi;
        sigma[k] := sin(a);
        omega[k] := cos(a);
        if abs(omega[k]) < 1e-9 then
            omega[k] := 0.0;
    end;
    k := 0;
    l := 0;
    repeat
        k := k + 1;
        if (sigma[k] < 0) then
        begin
            l := l + 1;
            sigma[l] := sigma[k];
            omega[l] := omega[k];
        end;
    until k = 2 * n;
end ( Butt );

procedure Cheby(n:integer; eps :real; var sigma,omega:Xarray);
var v,posexp, negexp,U :real;
    k,l :integer;
begin
    v := (1/n) * ln((1/eps) + sqrt(sqr(1/eps) + 1));
    posexp := exp(v) + exp(-v);
    negexp := exp(v) - exp(-v);
    for k := 1 to 2 * n do
    begin
        U := ((2 * k - 1)/(2 * n)) * pi;
        omega[k] := cos(U) * posexp / 2;
        if abs(omega[k]) < 1e-9 then
```

This procedure can be improved by having a look at the previous program.

```pascal
            omega[k] := 0.0;
            sigma[k] := sin(U) * negexp / 2;
        end;
    k := 0;
    l := 0;
    repeat
        k := k + 1;
        if sigma[k] < 0   then
        begin
            l := l + 1;
            sigma[l] := sigma[k];
            omega[l] := omega[k];
        end;
    until k = 2 * n;
end { Cheby };

procedure inputpoles;
var FilterType :char;
    dB,eps :real;
    i :integer;
begin
    nl(2);
    write('Enter number of active stages = ');
    readln(n);
    repeat
        write('Enter type of BP filter B(utterworth, C(hebyshev, or O(wn =');
        readln(FilterType);
    until FilterType in ['b','c','o'];
    case FilterType of
        'b': Butt(n,relp,imlp);
        'c': begin
                write('Enter ripple magnitude in dB = ');
                readln(dB);
                eps := sqrt((exp((dB/10)*ln(10))) - 1);
                Cheby(n,eps,relp,imlp)
             end;
        'o': begin
                for i := 1 to n do
                begin
                    write('Enter real and imaginary part of ');
                    write(i:1,' pole = ');
                    readln(relp[i],imlp[i]);
                end;
             end;
    end; { case filtertype };
end; { inputpoles }

procedure freqspecs;
var f1,f2,w1,w2 : real;
begin
    nl(2);
    write('Enter lower and upper frequency in Hz = ');
    readln(f1,f2);
    w1 :=2 * pi * f1;
    w2 :=2 * pi * f2;
    WC := sqrt(w2 * w1);
    BW := w2-w1;
end { freqspecs };

procedure printLPpoles;
var i :integer;
begin
    space(22);
    writeln('The Low Pass poles are ');
    nl(1);
    space(14);write('Real Part');space(17);writeln('Imag Part');
    nl(1);
    for i := 1 to n do
    begin
        space(8);
        write(relp[i]);
        space(4);
        writeln(imlp[i]);
    end;
end {printLPpoles };
```

```pascal
                procedure LPtoBP;
                var gama, riza, thet :Xarray;
                    i,k :integer;
                begin
                    for i := 1 to n do
                    begin
                        riza[i] := sqr(relp[i]) * sqr(BW) - (sqr(imlp[i])*sqr(BW)) - 4*sqr(WC);
                        thet[i] := (arctan(2*relp[i]*imlp[i]*sqr(BW)/riza[i]))/2;
                        gama[i] := sqrt(sqrt(sqr(riza[i])+sqr(2*relp[i]*imlp[i]*sqr(BW)*2)));
                    end;
                    for i := 1 to n do
                    begin
                        rebp1[i] := (relp[i] * BW + gama[i] * sin(thet[i])) / 2;
                        rebp2[i] := (relp[i] * BW - gama[i] * sin(thet[i])) / 2;
                        imbp1[i] := (imlp[i] * BW - gama[i] * cos(thet[i])) / 2;
                        imbp2[i] := (imlp[i] * BW + gama[i] * cos(thet[i])) / 2;
                    end;
                    nl(2);
                    space(21);
                    writeln('The Band Pass poles are ');
                    nl(1);space(14);write('Real Part');space(17);writeln('Imag Part');
                    nl(1);
                    for i := 1 to n do
                    begin
                        space(8);writeln(rebp1[i],' ':4,imbp1[i]);
                        space(8);writeln(rebp2[i],' ':4,imbp2[i]);
                    end;
                    nl(3);
                    writeln('there are ',n:1,' zeros at the origin due to the transformation');
                    nl(2);space(5);
                    writeln('The quadratic equations of the denominator are');
                    nl(2);
                    for k := 1 to n do
                    begin
                        i := k;
                        bp1[i] := -2 * rebp1[i];
                        bp2[i] := -2 * rebp2[i];
                        cp1[i] := sqr(rebp1[i]) + sqr(imbp1[i]);
                        cp2[i] := sqr(rebp2[i]) + sqr(imbp2[i]);
                        writeln('  (s**2 + ',bp1[i]:6:0,'s + ',cp1[i]:8:0,')');
                        i := i + 2;
                    end;
                end { LPtoBP };

                procedure ComponentCalculation;
                var R11,R22,R21,R12 :real;
                    C1,C2,Qfactor,reson :Xarray;
                    i,j :integer;
                begin
                    nl(2);
                    for i := 1 to n do
                    begin
                        write(' Enter C1 and C2 (in Farads) for ',i:1,' stage = ');
                        readln(C1[i],C2[i]);
                    end;
                    nl(3);
                    space(16);
                    writeln('Element values for second order stages ');nl(2);
                    space(9);write('R1');space(15);write('R2');space(15);
                    write('C1');space(15);writeln('C2');nl(2);
                    j := 1;
                    i := 1;
                    repeat
                        if rebp1[i] <> rebp2[i] then
                        begin
                            R21 := 1/(C2[j] * bp1[i]);
                            R11 := bp1[i] / (cp1[i] * C1[j]);
                            R12 := bp2[i] / (cp2[i] * C1[j+1]);
                            R22 := 1/(C2[j+1] * bp2[i]);
                            reson[j] := (1/sqrt(R21*R11*C1[j]*C2[j]))/(2*pi);
                            Qfactor[j] := sqrt((R11*C1[j])/(R21*C2[j]));
                            space(5);write(R11:8);space(9);write(R21:8);space(9);
                            write(C1[j]:8);space(9);writeln(C2[j]:8);
                            space(5);write(R12:8);space(9);write(R22:8);space(9);
                            write(C1[j+1]:8);space(9);writeln(C2[j+1]:8);
```

Low Pass to Band Pass transformation doubles the number of low pass poles.

Coefficient matching.

```
                Qfactor[j+1] := sqrt((R12*C1[j+1])/(R22*C2[j+1]));
                j := j + 2;
                i := i + 1;
            end
            else begin
                R21 := 1/(C2[j] * bp1[i]);
                R11 := bp1[i] / (cp1[i] * C1[j]);
                reson[j] := (1/sqrt(R21*R11*C1[j]*C2[j]))/(2*pi);
                Qfactor[j] := sqrt((R11*C1[j])/(R21*C2[j]));
                space(5);write(R11:8);space(9);write(R21:8);space(9);
                write(C1[j]:8);space(9);writeln(C2[j]:8);
                j := j + 1;
                i := i + 1;
            end;
        until i = n;
        nl(3);
        writeln('The resonant frequency and Q-factor of each stage are');
        nl(2);space(4);write('Section');space(5);write('Frequency');
        space(5);writeln('Q factor');nl(1);
        for i := 1 to n do
            writeln(' ':6,i:1,' ':9,reson[i]:8,' ':6,(1/Qfactor[i]):8);
end { component calculation };

begin
    inputpoles;
    printLPpoles;
    freqspecs;
    LPtoBP;
    ComponentCalculation;
end.
```

Execution begins...

Enter number of active stages = 3
Enter type of BP filter B(utterworth, C(hebyshev, or O(wn =c
Enter ripple magnitude in dB = 3.0
                The Low Pass poles are

        Real Part   Imag Part

    -1.49310104135098e-01   -9.03814428736114e-01
    -2.98620208270674e-01   0.00000000000000e+00
    -1.49310104135713e-01   9.03814428734873e-01

Enter lower and upper frequency in Hz = 1000.0 3000.0

                The Band Pass poles are

        Real Part   Imag Part

    -1.37477875422798e+03   -1.79719934084433e+04
    -5.01507350801746e+02   6.61432633034284e+03
    -1.87628610503274e+03   -1.07198322405285e+04
    -1.87628610503274e+03   1.07198322405285e+04
    -5.01507350804264e+02   -6.61432633034707e+03
    -1.37477875423320e+03   1.79719934084319e+04

there are 3 zeros at the origin due to the transformation

    The quadratic equations of the denominator are

(s**2 +   2750s + 324882564)
(s**2 +   3753s + 118435253)
(s**2 +   1003s + 44000822)

149

```
Enter C1 and C2 (in Farads) for 1 stage = 4.7e-9 4.7e-9
Enter C1 and C2 (in Farads) for 2 stage = 4.7e-9 4.7e-9
Enter C1 and C2 (in Farads) for 3 stage = 4.7e-9 4.7e-9
```

Element values for second order stages

R1	R2	C1	C2
1.8e+03	7.7e+04	4.7e-09	4.7e-09
4.9e+03	2.1e+05	4.7e-09	4.7e-09
6.7e+03	5.7e+04	4.7e-09	4.7e-09

The resonant frequency and Q-factor of each stage are

Section	Frequency	Q factor
1	2.9e+03	6.6e+00
2	0.0e+00	6.6e+00
3	1.7e+03	2.9e+00

Execution terminated.

Linear Passive Circuits

Fidler, J.K. and Nightingale, C. *Computer Aided Circuit Analysis.*

There are two basic methods that are used in the analysis of linear passive circuits namely loop analysis and nodal analysis. Using nodal analysis, the nodal equations of an n-node passive circuit can be written as:

$$I_j = \sum_{k=1}^{n} y_{jk} V_k \qquad (7.13)$$

or in matrix form as

$$\begin{bmatrix} I_1 \\ I_2 \\ I_3 \\ \cdot \\ \cdot \\ \cdot \\ I_n \end{bmatrix} = \begin{bmatrix} y_{11} & y_{12} & y_{13} & \cdots & y_{1n} \\ y_{21} & y_{22} & y_{23} & \cdots & y_{2n} \\ y_{31} & y_{32} & y_{33} & \cdots & y_{3n} \\ \cdot & \cdot & \cdot & & \cdot \\ \cdot & \cdot & \cdot & & \cdot \\ \cdot & \cdot & \cdot & & \cdot \\ y_{n1} & y_{n2} & y_{n3} & & y_{nn} \end{bmatrix} \begin{bmatrix} V_1 \\ V_2 \\ V_3 \\ \cdot \\ \cdot \\ \cdot \\ V_n \end{bmatrix} \qquad (7.14)$$

where y_{jj} = the sum of all admittances connected to node j
and y_{jk} = the negative sum of all admittances connected between node j and node k.

The nodes are numbered 1....n with the reference node as node 0. The circuit can be analysed by solving the simultaneous equations given by Equation 7.14. There are a number of methods which can be used to solve these equations, such as Gaussian elimination, L-U factorisation and pivotal condensation. In the program given in the next section the pivotal condensation technique is used to condense a given n-terminal network to a two port. This can be achieved by reducing the order of the (nxn) matrix by one until a (2 × 2) matrix is obtained. In effect, one node is removed at a time from the circuit until a two port network is obtained. Every time the matrix is reduced by one order the information contained in the rows and columns to be deleted is transferred to the remaining ones. In general, in an (nxn) nodal admittance matrix, node k may be suppressed from the circuit by selecting as pivot the matrix element y_{kk}. This pivot can then be used to replace every matrix element y_{ij} according to

$$y_{ij} = y_{ij} - \frac{y_{ik} y_{kj}}{y_{kk}} \tag{7.15}$$

followed by the deletion of the kth row and column.

From the resulting two port network the short-circuit input and output admittances can easily be calculated as well as the voltage and current transfer ratios.

Circuit Analysis Program

```
program RLCcircuitAnalysis(input, output);

type sub = 1 .. 30;
     nodes = 0 .. 29;
     matrix = array[sub,sub] of real;
     topol = array[nodes,nodes] of real;
     twoport = array [1..2,1..2] of real;

var elvalue, re, im, f1, f2, dc :real;
    omega, mag, freqstep, phase, logstep :real;
    maxnodes, branches, npoints,i,j :integer;
    el, hzrad, linlog, ratdB, Degra, YN :char;
    r,l,c :topol;
    gadmatrix, hadmatrix, cadmatrix :matrix;
    g, x :twoport;

function accept(test,c1,c2 :char):boolean;
begin
    accept := test in [c1,c2]
end { accept };

procedure compmatrix(n1,n2:nodes;value:real;var Admatrix :matrix);
begin
    if n1 = n2 then
        writeln('Short circuit element ... it will be ignored')
    else
        if (n1=0) or (n2=0) then
            if n1>n2 then
                Admatrix[n1,n1] := Admatrix[n1,n1] + value
            else
                Admatrix[n2,n2] := Admatrix[n2,n2] + value
        else begin
            Admatrix[n1,n2] := Admatrix[n1,n2] - value;
            Admatrix[n2,n1] := Admatrix[n2,n1];
            Admatrix[n1,n1] := Admatrix[n1,n1] + value;
            Admatrix[n2,n2] := Admatrix[n2,n2] + value
        end
end { procedure compmatrix };

procedure compmult(a,b,c,d:real; var re,im :real);
begin
    re := (a*c) - (b*d);
    im := (b*c) + (a*d)
end { procedure compmult };

procedure compdiv(a,b,c,d:real; var ree,imm :real);
begin
    ree := ((a*c) + (b*d))/(c*c + d*d);
    imm := ((c*b) - (a*d))/(c*c + d*d);
end { procedure compdiv };
```

```pascal
procedure pivotcond(maxnodes:nodes; omega:real; c,H,G:matrix; var g,x:twoport);
var i,j,k,l :nodes;
    re,im,ree,imm :real;
    rA,IA :matrix;
begin
    for i := 1 to maxnodes do
        for j := 1 to maxnodes do
        begin
            IA[i,j] := c[i,j] * omega + H[i,j] / omega;
            rA[i,j] := G[i,j];
        end;
    for k := (maxnodes -1) downto 2 do
    begin
        for i := 1 to maxnodes do
            for j := 1 to maxnodes do
            begin
                if ((i <> k) and (j <> k)) then
                begin
                    compmult(rA[i,k],IA[i,k],rA[k,j],IA[k,j],re,im);
                    compdiv(re,im,rA[k,k],IA[k,k],ree,imm);
                    rA[i,j] := rA[i,j] - ree;
                    IA[i,j] := IA[i,j] - imm;
                end
            end;
            for l := 1 to maxnodes do
            begin
                rA[l,k] := 0.0;
                IA[l,k] := 0.0;
                rA[k,l] := 0.0;
                IA[k,l] := 0.0
            end;
    end;
    g[1,1] := rA[1,1];
    x[1,1] := IA[1,1];
    g[1,2] := rA[1,maxnodes];
    x[1,2] := IA[1,maxnodes];
    g[2,1] := rA[maxnodes,1];
    x[2,1] := IA[maxnodes,1];
    g[2,2] := rA[maxnodes,maxnodes];
    x[2,2] := IA[maxnodes,maxnodes]
end; { procedure pivotcond }

function atan(x,y:real; Degrad:char):real;
const pi = 3.141492653589;
var phi,efap :real;
begin
    if x = 0.0 then
        if y > 0.0 then
            efap := pi/2
        else
            efap := -pi/2;
    if y = 0.0 then
        if x >= 0.0 then
            efap := 0.0
        else
            efap := pi;
    if ((x <> 0.0) and (y <> 0.0)) then
    begin
        phi := arctan(abs(y/x));
        if x > 0.0 then
            if y > 0.0 then
                efap := phi
            else
                efap := -phi
        else
            if y > 0.0 then
                efap := pi - phi
            else
                efap := -pi + phi
    end;
    if Degrad = 'D' then
        atan := efap * 180 / pi
    else
        atan := efap
end { function atan };
```

```
procedure readcomponents;
var i,j,k :integer;
begin
    writeln;writeln;
    writeln('**** circuit Specification ****');
    write('Number of branches = ');readln(branches);
    write('Number of nodes    = ');readln(maxnodes);
    writeln('The components should be entered in the following format');
    writeln(' ':11,'<type> <node> <node> <value>');
    for k := 1 to branches do
    begin
        repeat
            write('branch',k:2,' = ');
            readln(el,i,j,elvalue);
        until (el in ['r','c','l']) and (i<=maxnodes) and (elvalue>0) and (j<=ma
        case el of
            'c' :begin
                    c[i,j] := elvalue;
                    compmatrix(i,j,elvalue,cadmatrix)
                 end;
            'l' :begin
                    l[i,j] := elvalue;
                    compmatrix(i,j,-1/elvalue,hadmatrix)
                 end;
            'r' :begin
                    r[i,j] := elvalue;
                    compmatrix(i,j,1/elvalue,gadmatrix)
                 end
        end { case element }
    end
end { readcomponents };

procedure writecomponents;
var i,j :nodes;
begin
    writeln;writeln;
    writeln('component type       from     to       value');
    for i := 0 to maxnodes do
        for j := 0 to maxnodes do
        begin
            if r[i,j] > 0 then
                writeln('resistance',' ':7,i:6,j:6,' ':4,r[i,j]:1:3,' Ohms');
            if l[i,j] > 0 then
                writeln('Inductance',' ':7,i:6,j:6,' ':4,l[i,j]:1:3,' Henries');
            if c[i,j] > 0 then
                writeln('Capacitance',' ':6,i:6,j:6,' ':4,c[i,j]:1:3,' Farads');
        end;
end; { procedure writecomponents }

procedure freqspecs;
var dummy :real;
begin
    writeln;writeln;
    writeln('  ****    Frequency response Specifications    ****');
    writeln('  --------------------------------');
    writeln;writeln;
    repeat
        write('    Response in Hz or rad/s ? enter H or r = ');
        readln(hzrad);
    until accept(hzrad,'H','r');
    repeat
        write('    Enter lower and upper frequency limits in ');
        if hzrad = 'H' then
            write('Hz = ')
        else
            write('rad/s = ');
        readln(f1,f2);
    until ((f1>=0) and (f2>0));
    if f1 > f2 then
    begin
        dummy := f2;
        f2 := f1;
        f1 := dummy
    end;
```

```
        repeat
            write('    Frequency response in linear or log form (enter l or G) = ');
            readln(linlog);
        until accept(linlog,'l','G');
        if linlog = 'l' then
        begin
            write('      Enter Frequency step in ');
            if hzrad = 'H' then
                write('Hz = ')
            else
                write('rad/s = ');
            readln(freqstep);
            if f1 = 0.0 then
                f1 := 1e-9;
            npoints := round((f2-f1)/freqstep)+1
        end else begin
            write(' ':5);
            write('Number of points for log frequency scale = ');
            readln(npoints);
            if (f1 = 0.0) then
                if (f2 > 1.000001) then
                    f1 := 1/f2
                else
                    f1 := 1e-3;
            dummy := (ln(f2/f1)/ln(10))/(npoints - 1);
            logstep := exp(dummy*ln(10))
        end;
        repeat
            write('     Output as a ratio or in dB (enter r or d) = ');
            readln(ratdB);
        until accept(ratdB,'r','d');
        repeat
            write('     Phase in degrees or radians (enter D or r) = ');
            readln(Degra);
        until accept(Degra,'r','D');
end; { freqspecs }

procedure calculate;
var i :integer;
begin
    repeat
        write('    Magnitude relative to zero ? y/n = ');
        readln(YN);
    until accept(YN,'y','n');
    if YN = 'y' then
    begin
        pivotcond(maxnodes,1e-9,cadmatrix,hadmatrix,gadmatrix,g,x);
        compdiv(-g[2,1],-x[2,1],g[2,2],x[2,2],re,im);
        dc := 1/(sqrt(re*re+im*im));
    end else dc := 1;
    writeln;writeln;
    writeln(' ':20,'Frequency response');
    writeln;
    writeln(' ':5,'Frequency',' ':13,'Magnitude',' ':14,'Phase');
    if hzrad  = 'H' then
        write(' ':6,'  Hz ')
    else
        write(' ':6,'rad/s');
    if ratdB = 'd' then
        write(' ':18,' dB  ')
    else
        write(' ':18,'ratio');
    if Degra = 'D' then
        writeln(' ':16,'degrees')
    else
        writeln(' ':16,'radians');
    omega := f1;
    for i := 1 to npoints do
    begin
        pivotcond(maxnodes,omega,cadmatrix,hadmatrix,gadmatrix,g,x);
        compdiv(-g[2,1],-x[2,1],g[2,2],x[2,2],re,im);
        if ratdB = 'r' then
            mag := dc * sqrt(sqr(re)+sqr(im))
        else
            mag := 20*ln(dc*sqrt(sqr(re)+sqr(im)))/ln(10);
```

```
            if Degra = 'D' then
                phase := atan(re,im,'D')
            else
                phase := atan(re,im,'R');          (flag this for change)
            write(' ':1,omega:10:3,' ':13,mag:10:3);
            writeln(' ':12,phase:10:3);
            if linlog = 'I' then
                omega := omega + freqstep
            else
                omega := omega * logstep;
        end
end { calculate };

{main program starts here}
begin
    writeln;writeln;
    readcomponents;
    writecomponents;
    repeat
        freqspecs;
        calculate;
        writeln;writeln;
        write(' ? rerun circuit with different frequency specs y/n = ');
        readln(YN);
    until accept(YN,'y','n')
end.
```

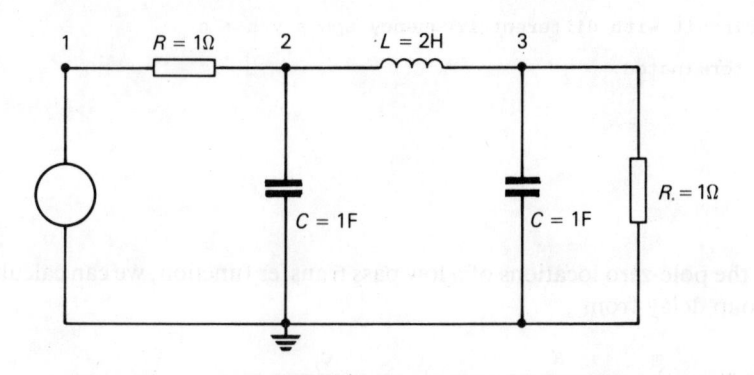

Fig. 7.3

```
Execution begins...

**** circuit Specification ****

Number of branches = 5
Number of nodes    = 3
The components should be entered in the following format
            <type> <node> <node> <value>
    branch 1 = r 1 2 1
    branch 2 = c 2 0 1
    branch 3 = l 2 3 2
    branch 4 = c 3 0 1
    branch 5 = r 3 0 1


component type         from      to         value
resistance              1        2       1.000 Ohms
Capacitance             2        0       1.000 Farads
Inductance              2        3       2.000 Henries
resistance              3        0       1.000 Ohms
Capacitance             3        0       1.000 Farads
```

```
****    Frequency response Specifications    ****
        ---------------------------------

        Response in Hz or rad/s ? enter H or r = r
        Enter lower and upper frequency limits in rad/s = 0 10.0
        Frequency response in linear or log form (enter I or G) = G
        Number of points for log frequency scale = 10
        Output as a ratio or in dB (enter r or d) = d
        Phase in degrees or radians (enter D or r) = D
        Magnitude relative to zero ? y/n = y

                        Frequency response

        Frequency           Magnitude              Phase
        rad/s                  dB                 degrees
        0.100                -0.000              -11.479
        0.167                -0.000              -19.207
        0.278                -0.002              -32.334
        0.464                -0.043              -55.510
        0.774                -0.847             -100.395
        1.292                -7.514             -169.602
        2.154               -20.043              145.507
        3.594               -33.335              122.331
        5.995               -46.667              109.204
       10.000               -60.000              101.476

    ? rerun circuit with different frequency specs y/n = n

    Execution terminated.
```

Problems

7.1 Given the pole-zero locations of a low pass transfer function, we can calculate the group delay from

$$\tau(\omega) = \sum_{i=1}^{m} \frac{a_i}{a_i^2 + (\omega + b_i)^2} - \sum_{j=1}^{n} \frac{c_j}{c_j^2 + (\omega + d_j)^2}$$

where

$a_i + jb_i$ = coordinates of a zero
$c_j + jd_j$ = coordinates of a pole

Write a procedure to calculate the group delay and hence incorporate this in the TransferFunctionAnalysis program.

7.2 Write a plotting routine and hence add it to the TransferFunctionAnalysis program. You should give the option to the user for graphical or tabular results for any required response.

7.3 From the circuit shown in Fig. 7.2 the low pass characteristic can be obtained:

$$\frac{V_2}{V_1}(s) = \frac{\frac{1}{C_1 C_2 R_1 R_2}}{s^2 + s\frac{1}{R_2 C_2} + \frac{1}{C_1 C_2 R_1 R_2}}$$

With this knowledge, calculate the element values of a Butterworth low pass filter with cut-off frequency = 1000 kHz.

7.4 For high Q-factor values there is a loading effect between stages using the circuit shown in Fig. 7.2. To overcome this problem the R_1 resistor in each second order stage can be replaced by a voltage dividing circuit as shown.

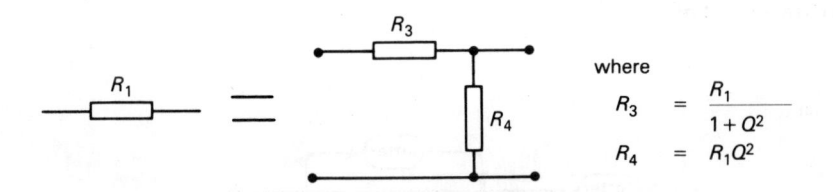

where
$$R_3 = \frac{R_1}{1+Q^2}$$
$$R_4 = R_1 Q^2$$

Incorporate this circuit modification in your program.

7.5 Add a plotting procedure to the ActiveFilterDesign program and plot the magnitude response of the specified active filter.

7.6 Modify the RLCcircuitAnalysis program to calculate the short circuit current ratio $\dfrac{I_2}{I_1}$

7.7 Modify the RLCcircuitAnalysis program to calculate the input and output impedance of the circuit.

Appendix A

Syntax Diagrams

identifier

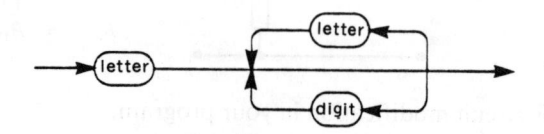

simple type

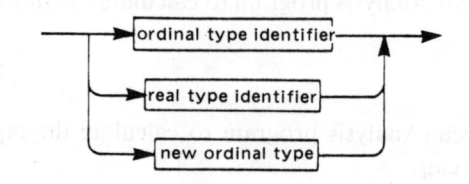

- ordinal type identifier
- real type identifier
- new ordinal type

new ordinal type

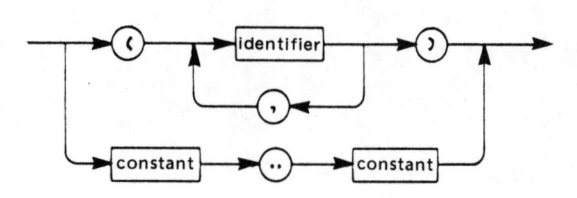

(identifier , ...)
constant .. constant

new structured type

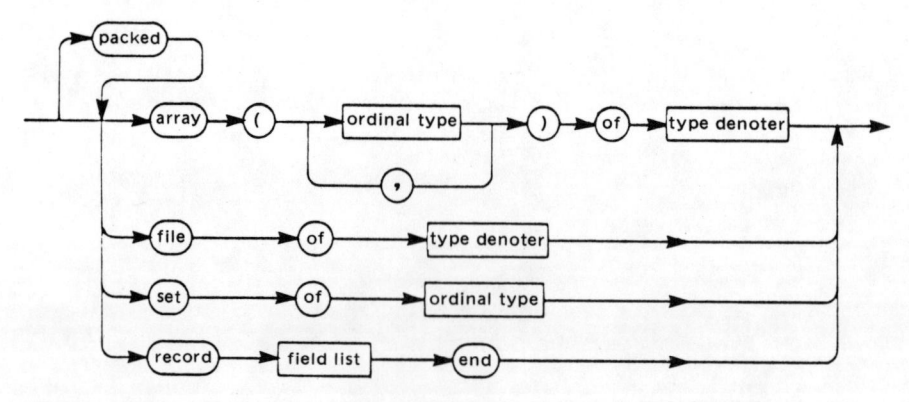

- packed
- array (ordinal type , ...) of type denoter
- file of type denoter
- set of ordinal type
- record field list end

type denoter

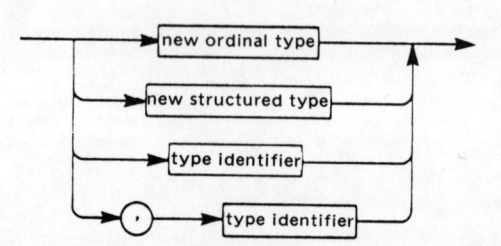

- new ordinal type
- new structured type
- type identifier
- , type identifier

procedure heading

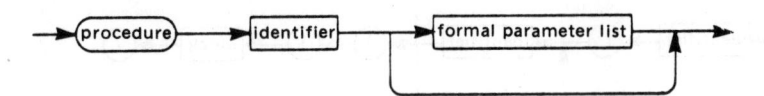

```
──▶(procedure)──▶[identifier]──┬──▶[formal parameter list]──┬──▶
                               │                            │
                               └────────────────────────────┘
```

procedure declaration

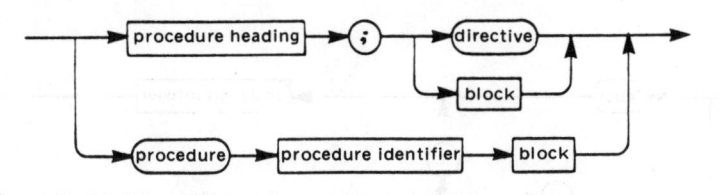

```
────┬──▶[procedure heading]──▶(;)──┬──▶(directive)──┬──▶
    │                              └──▶[block]──────┘
    │
    └──▶(procedure)──▶[procedure identifier]──▶[block]──▶
```

function heading

```
──▶(function)──▶[identifier]──┬──▶[formal parameter list]──┬──▶(:)──┬──▶[simple type identifier]──┬──▶
                              │                            │        └──▶[pointer type identifier]─┘
                              └────────────────────────────┘
```

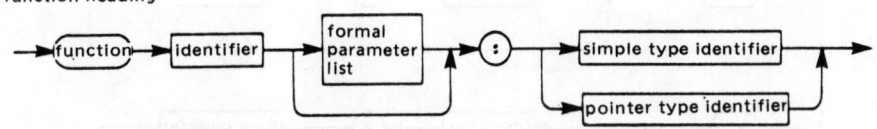

function declaration

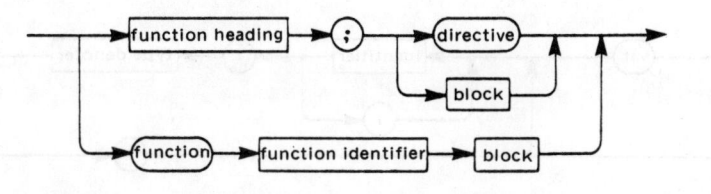

```
────┬──▶[function heading]──▶(;)──┬──▶(directive)──┬──▶
    │                             └──▶[block]──────┘
    │
    └──▶(function)──▶[function identifier]──▶[block]──▶
```

formal parameter section

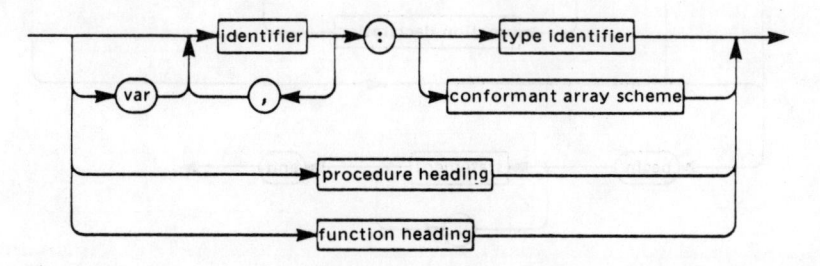

```
────┬──┬──────────┬──▶[identifier]──▶(:)──┬──▶[type identifier]─────────┬──▶
    │  └──(var)───┘      ▲     │          └──▶[conformant array scheme]─┘
    │                    └─(,)─┘
    │
    ├──▶[procedure heading]──────────────────────────────────────▶
    │
    └──▶[function heading]───────────────────────────────────────▶
```

program

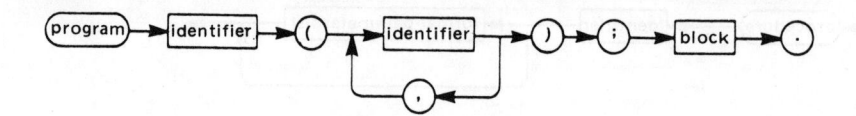

block

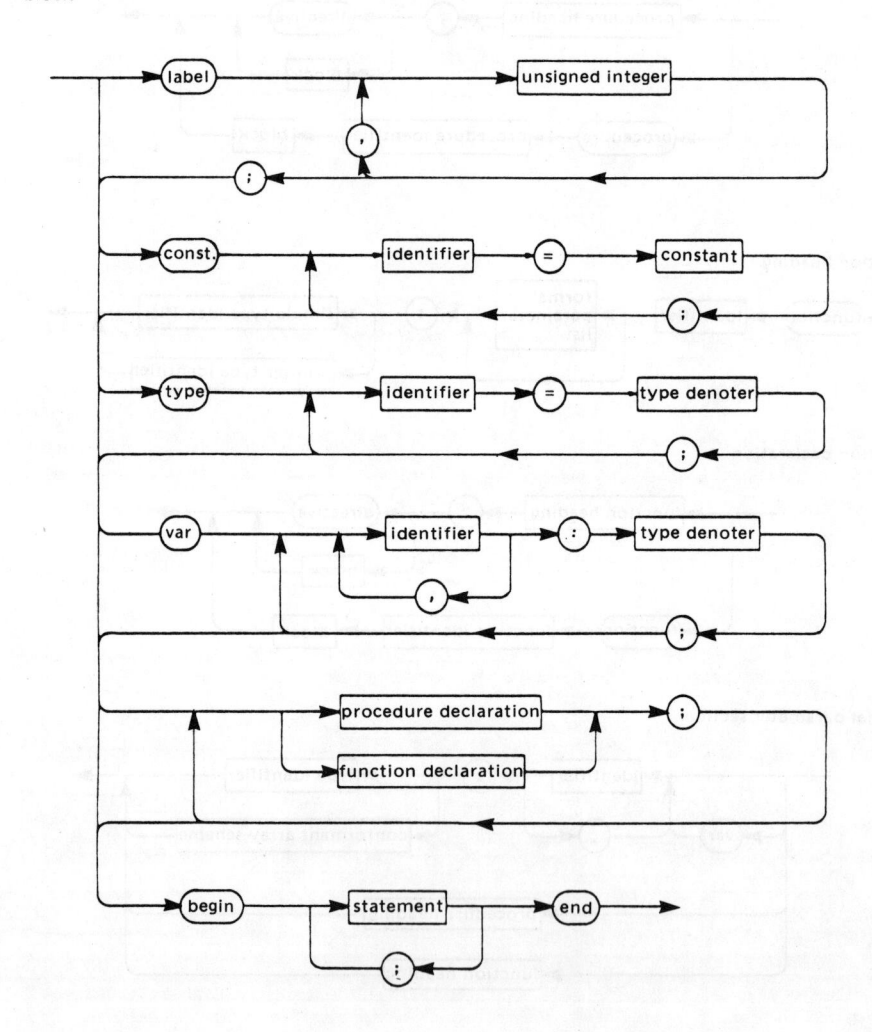

statement

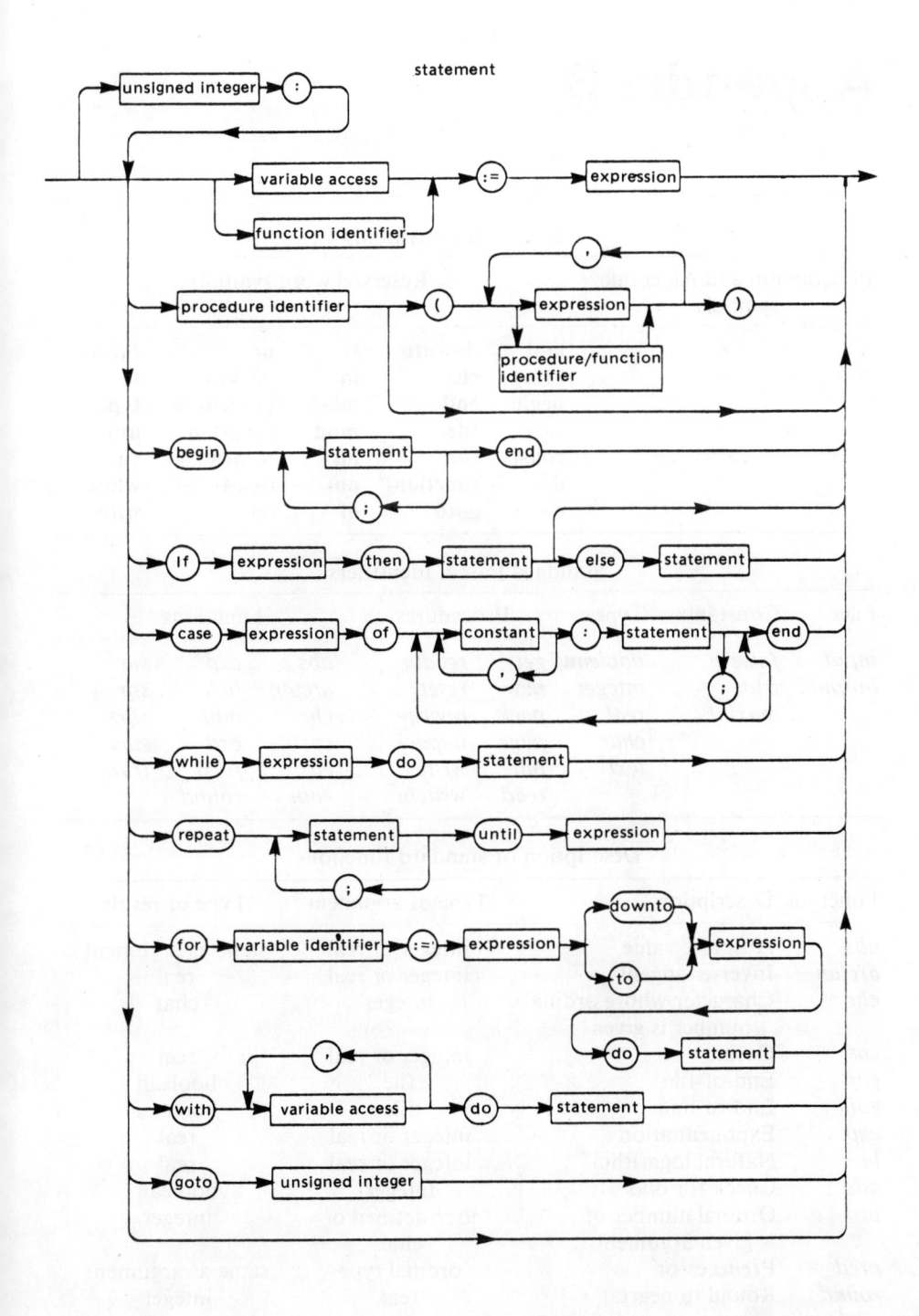

Appendix B

Pascal Special Symbols

Punctuation and Algebraic symbols	Reserved word symbols

```
  +    −    *    /         and      downto    if      or         then
  >    <    [    ]         array    else      in      packed     to
  ;    ,    :    .         begin    end       label   procedure  type
  ^    (    )    ..        case     file      mod     program    until
  <>   <=   >=   :=        const    for       nil     record     var
  =    {    }              div      function  not     repeat     while
                           do       goto      of      set        with
```

Standard Pascal Identifiers

Files	Constants	Types	Procedures		Functions		
input	*false*	*boolean*	*get*	*readln*	*abs*	*exp*	*sin*
output	*true*	*integer*	*new*	*reset*	*arctan*	*ln*	*sqr*
	maxint	*real*	*pack*	*rewrite*	*chr*	*odd*	*sqrt*
		char	*page*	*unpack*	*cos*	*ord*	*succ*
		text	*put*	*write*	*eof*	*pred*	*trunc*
			read	*writeln*	*eoln*	*round*	

Description of standard functions

Function	Description	Type of argument	Type of result
abs	Absolute value	integer or real	same as argument
arctan	Inverse tangent	integer or real	real
chr	Character whose ordinal number is given	integer	char
cos	Cosine	integer or real	real
eof	End-of-file	file	boolean
eoln	End-of-line	file	boolean
exp	Exponentiation	integer or real	real
ln	Natural logarithm	integer or real	real
odd	Check for odd	integer	boolean
ord	Ordinal number of given argument	user-defined or char	integer
pred	Predecessor	ordinal type	same as argument
round	Round to nearest integer	real	integer
sin	Sine	integer or real	real
sqr	Square	integer or real	same as argument
sqrt	Square-root	integer or real	real
succ	Successor	ordinal type	same as argument
trunc	Truncate	real	integer

References

1. Wilson, I.R. and Addyman, A.M. *A Practical Introduction to Pascal-with BS6192* (MacMillan, 1982).
2. Welsh, J. and Elder, J. *Introduction to Pascal* (Prentice-Hall, 1982).
3. Atkinson, L.V. and Harley, P.J. *An Introduction to Numerical Methods with Pascal* (Addison-Wesley, 1983).
4. Brown, P.J. *Pascal from BASIC* (Addison Wesley 1982).
5. Graham, N. *Introduction to Pascal* (West Publishing Company, 1980).
6. Cooper, D. and Clancy, M. *Oh! Pascal!* (Norton, W.W. & Company, 1982).
7. Schneider, G.M., Weingart, S.W. and Perlman, D.M. *An Introduction to Programming and Problem Solving with Pascal* (Wiley, J. & Sons, 1978).
8. Jensen, K. and Wirth, N. *Pascal: User Manual and Report* (Springer-Verlag, 1975).
9. Findlay, W. and Watt, D.A. *Pascal: An Introduction to Methodical Programming* (Pitman, 1981).
10. Atkinson, L.V. *Pascal Programming* (Wiley, J. & Sons, 1980).
11. Ledgard, H.F., Hueras, J.F. and Nagin, P.A. *Pascal with Style* (Hayden,1979).
12. Grogono, P. *Programming in Pascal* (Addison-Wesley, 1978).

Engineering References

1. Fidler, J.K. *Introductory Circuit Theory* (McGraw Hill, 1980).
2. Attikiouzel, J. and Linggard, R. Low Cost Active Low Pass and Band Pass Network, *International Journal of Circuit Theory and Applications* Vol. 2, No. 4, pp. 397–400, 1974.
3. Ritchie, G.J. *Transistor Circuit Techniques*, Tutorial Guides in Electronic Engineering (Van Nostrand Reinhold, 1983).
4. Horrocks, D.H. *Feedback Circuits and Op.Amps*, Tutorial Guides in Electronic Engineering (Van Nostrand Reinhold, 1983).
5. Kuo, F.F. *Network Analysis and Synthesis* (Wiley, J. & Sons 1966).
6. Fidler, J.K. and Nightingale, C. *Computer Aided Circuit Design* (Nelson, 1978).
7. Distefano, J.J., Stubberub, A.R. and Williams, I.J. *Feedback and Control Systems*, Schaum's Outline Series (McGraw Hill, 1967).
8. Dietmeyer, D.L. *Logic Design of Digital Systems* (Allyn and Bacon Inc., 1978).
9. Zverev, A. *Handbook of Filter Synthesis* (Wiley, J. & Sons, 1967).
10. Temes, G.C. and LaPatra, J.W. *Circuit Synthesis and Design* (McGraw-Hill, 1977).
11. Lathi, B.P. *Signal, Systems and Communications* (Wiley, J. & Sons, 1965).

Index

Abs 20
Accuracy of reals 17
Active filters 144
Actual parameter 78
Algebraic symbols 5
ALGOL 2
Algorithm 1
And 37
Arctan 20, 51
Arithmetic operators 18, 20
Array 43, 91, 107
ASCII 32
Assembly language 2
Assignment 8, 47

BASIC 2
Begin 8
Binary digit 2
Block 7
Block structure 86
Boolean 37
Boundary conditions 37

C 2
Cardinality 39
Case 52
Char 33
Character set 32
Chr 35
Circumflex 132
COBOL 2
Collating sequence 32
Comment 9
Compiler 3
Compound statement 48
Conditional statement 48
Const 14
Continuation test 54
Control character 62

Data structures 5
Data type 4
Debugging 3
Declarations 9, 15
Definition part 9
Directive 103
Dispose 132
Div 18, 38
Do 54, 61
Downto 61
Dummy parameter 78, 90

E 5
EBCDIC 33
Editor 3
Else 48
Empty set 120
Empty statement 48
End 8
Enumerated type 40
Eof 38, 116
Eoln 38, 117
Error 3
Extended Backus-Naur Form 8
External files 114

Factorial 102
False 37
Field width 23, 35
File 113
Fixed point notation 16
Floating point error 69
Floating point notation 16
Formal parameter 78, 90
Format, output 23
For statement 61
Forward directive 103
Fourier 66
Function 78

Get 115
Global identifier 82, 93
Goto 67
Graph 70, 91

Heading 7
Heap 132
Hierarchy, operator 38
High-level language 2

Identifier 5
If 48
Implementation 3
In 122
Input 7, 21, 34
Integer 16
Internal files 114
Inverse junction 35
Iterative 102

Label 67
Label, case 52

165

Limitation, set-size 120
Ln 20
Listing 3
Local identifier 82, 87
Logical operators 38

Machine language 2
Maxint 16
Memory 3
Mnemonic 5
Mod 18
Modular program 78
Mutual recursion 102

Nesting 49, 63
New 132
Nil 133

Odd 38, 80
Or 37
Ord 35
Ordinal type 5, 32, 39
Output 7, 22, 34
Overflow 69

Pack 110
Packed array 110
Page 23
Parent 42
Pascal 4
Pointer 4, 131
Precedence 38
Pred 20, 35
Procedure 89
Put 115

Read 21
Readln 21
Real 16
Record 123
Recursion 102
Relational operators 34, 120
Repeat-until 58
Reserved word 5
Reset 115
Rewrite 115
Robust 40
Round 20

Run 3
Run-time error 3

Scalar data type 4, 39
Scope 86
Semantic 3
Sequential 113
Set 119
Short-hand notation 40, 43
Side-effects 82, 86
Simple type 4
Software tools 3
Stack 131
Standard identifier 6
String 111
Structured data type 4, 107
Subprogram 77
Subrange 41
Subscript 43
Succ 20, 35
Syntax 6, 158–161
Syntax error 3

Text 116, 117
Textfile 116
Then 48
To 61
Top-down 78
Transfer function 35, 39
Translator 3
True 37
Trunc 20
Type 40

Union 121
Unpack 111
Unstructured data type 4
Until 58
Up-arrow 132
User-defined 6

Value-parameter 90
Var 14
Variable parameter 94
Variant record 128

While statement 54
With statement 127
Write 22
Writeln 22